Math Bridge™
8th grade

Written by:

Tracy Dankberg & James Michael Orr

D0932420

Project Directors: Michele D. Van Leeuwen
Scott G. Van Leeuwen

Creative & Marketing Director: George Starks

Design & Technical Project Director: Dante J. Orazzi

TABLE OF CONTENTS

INTRODUCTION

The *Math Bridge* series is designed to help students improve their mathematical skills in all areas. This book has been developed to provide eighth grade students with skill-based exercises in the following areas: computation and estimation; decimals; number theory; rational numbers; ratio and proportion; percent; geometry; graphing; integers; equations and inequalities; probability and statistics; polynomials. The purpose of this book, therefore, is to strengthen students' mathematical concepts, thus helping them to become better mathematicians and to improve achievement test scores.

Math Bridge includes many extras to help your students in their study of mathematics. For instance,

✔ An Incentive Contract begins the book to motivate students to complete their work.

✔ A diagnostic test has been included to help assess your students' mathematical knowledge.

✔ Exercises become progressively more difficult as students work through the book.

✔ Tips are included throughout the book as reminders to help students successfully complete their work.

✔ Thought-provoking questions (Think About It) are periodically placed throughout the book to emphasize critical thinking skills.

✔ Additional exercises are included to help students in practicing with estimation.

✔ The exercises prepare students for standardized achievement tests.

✔ Each section includes problem-solving exercises written with the purpose of reinforcing the skills taught in that section.

Mathematics is all around us and is an essential part of life. It is the authors' intention that through the completion of this book, students will come away with a stronger knowledge of mathematics to assist them both inside and outside of the classroom.

Welcome to MATH FACTS

Math Facts, which are listed throughout the book, make mathematics more interesting by grounding it in world history, events, and science.

Incentive Contract

In • cen'tive, *n.* **1.** Something that urges a person on. **2.** Enticing. **3.** Encouraging. **4.** That which excites to action or moves the mind.

LIST YOUR AGREED-UPON INCENTIVE FOR EACH SECTION BELOW
Place a ✔ after each activity upon completion

Students Signature _____

PG	Activity Title	✔
9	Estimation Strategies	
10	Powers & Exponents	
11	More Exponents	
12	Order of Operations	
13	Problem Solving	

MY INCENTIVE IS ✗

PG	Activity Title	✔
14	Integers & Absolute Value	
15	Adding Integers	
16	Subtracting Integers	
17	Multiplying & Dividing	
18	Problem Solving	

MY INCENTIVE IS ✗

PG	Activity Title	✔
19	Writing Algebraic Expressions & Equations	
20	Evaluating Expressions	
21	Solving Addition & Subtraction Equations	
22	Solving Multiplication & Division Equations	
23	Solving Two-Step Equations	
24	Solving Inequalities	
25	Problem Solving	

MY INCENTIVE IS ✗

PG	Activity Title	✔
26	Comparing & Ordering	
27	Estimating	
28	Adding & Subtracting	
29	Multiplying & Dividing	
30	Solving Equations	
31	The Metric System	
32	Scientific Notation	
33	Problem Solving	

MY INCENTIVE IS ✗

PG	Activity Title	✔
34	Squares & Square Roots	
35	Multiples & Factors	
36	Prime & Composite Numbers	
37	Prime Factorization	
38	Greatest Common Factor	
39	Lowest Common Multiple	
40	Problem Solving	

MY INCENTIVE IS ✗

PG	Activity Title	✔
41	Simplifying Numbers	
42	Improper Fractions	
43	Comparing & Ordering	
44	Adding & Subtracting	
46	Solving Equations	
47	Multiplying Fractions & Mixed Numbers	
48	Dividing Fractions & Mixed Numbers	
49	Solving Equations	

MY INCENTIVE IS ✗

PG	Activity Title	✔
50	Ratio & Equal Ratios	
51	Proportions	
52	Rates	
53	Percent	
54	Percents & Decimals	

Teacher or Parent Signature _____

PG	Activity Title	✔
55	Problem Solving	

MY INCENTIVE IS ✗

PG	Activity Title	✔
56	Percent of a Number	
57	Simple Interest	
58	Percent of Increase or Decrease	
59	Problem Solving	

MY INCENTIVE IS ✗

PG	Activity Title	✔
60	Perimeter	
61	Angles	
62	Classifying Triangles	
63	Pythagorean Theorem	
64	Circumference & Area of Circles	
65	Area of Rectangles & Parallelograms	
66	Area of Triangles & Trapezoids	
67	Volume of Prisms & Cylinders	
68	Volume of Pyramids & Cones	
69	Surface Area	
70	Problem Solving	

MY INCENTIVE IS ✗

PG	Activity Title	✔
71	Fundamental Counting Principle	
72	Permutations	
73	Combinations	
74	Independent & Dependent Events	
75	Problem Solving	
76	Stem & Leaf Plots	
77	Histograms	
78	Line Plots	

MY INCENTIVE IS ✗

PG	Activity Title	✔
79	Graphing on a Coordinate Plane	
80	Graphing on a Number Line	
81	Graphing Linear Equations	

MY INCENTIVE IS ✗

PG	Activity Title	✔
82	Combining Like Terms	
83	Simplifying to Solve	
84	Variables on Both Sides	
85	Solving Multi-Step Inequalities	

MY INCENTIVE IS ✗

PG	Activity Title	✔
86	Introduction	
87	Adding Polynomials	
88	Subtracting Polynomials	
89	Multiplying Polynomials	
90	Dividing Polynomials	

MY INCENTIVE IS ✗

DIAGNOSTIC TEST

Name _____ **Score** _____

Directions: Read the following problems. For each question, fill in the circle of the correct answer. If the correct answer is not given, fill in the answer space marked N (Not Given).

1. $3 \times 8 - 6 \div 2 =$ ○ A. 3   ○ B. 9   ○ C. 13   ○ D. 21

2. $8 + 6 - 3 =$ ○ A. 6   ○ B. 17   ○ C. 26   ○ D. N

3. $4 \cdot 7 + 4 \cdot 3 =$ ○ A. 40   ○ B. 80   ○ C. 96   ○ D. 132

4. $-18 \cdot 3 =$ ○ A. –6   ○ B. 6   ○ C. –54   ○ D. 54

5. $24 \div (-8) =$ ○ A. –3   ○ B. 3   ○ C. –4   ○ D. N

6. $\dfrac{7 + (25 \div 5)}{4} =$ ○ A. 2   ○ B. 3   ○ C. 4   ○ D. 5

7. $-6 - 3 =$ ○ A. –3   ○ B. 3   ○ C. 2   ○ D. N

8. $-4 + 8 =$ ○ A. –2   ○ B. 4   ○ C. 12   ○ D. –4

9. $3 + (-10) =$ ○ A. 7   ○ B. –7   ○ C. 13   ○ D. –13

10. $6 \div 3 \times -2 =$ ○ A. –4   ○ B. –1   ○ C. 4   ○ D. 1

11. $-6.8 \times 3.2 =$ ○ A. 2.176   ○ B. 21.76   ○ C. 2.125   ○ D. N

12. $42.35 \div 12.1 =$ ○ A. 2.5   ○ B. 3.5   ○ C. 35   ○ D. N

13. $19.1 - (-6.7) =$ ○ A. 12.4   ○ B. 86.1   ○ C. 25.8   ○ D. –47.9

14. What is the closest estimate of $1215 \div 62$?
   ○ A. 20   ○ B. 25   ○ C. 200   ○ D. 250

15. The closest estimate of 85×205 is what?
   ○ A. 15,000   ○ B. 16,000   ○ C. 21,000   ○ D. 22,000

16. What is the closest estimate of $21.95 - 3.02$?
   ○ A. 16   ○ B. 17   ○ C. 18   ○ D. 19

17. The closest estimate of $\$17.18 + \$9.63$ is what?
   ○ A. $26.00   ○ B. $27.00   ○ C. $28.00   ○ D. $29.00

18. What is the closest estimate of $.49 (18)$?
   ○ A. 7   ○ B. 8   ○ C. 9   ○ D. N

Name _____

19. What is 6^3?

  ○ A. 18   ○ B. 36   ○ C. 216   ○ D. 1296

20. What is the greatest common factor of 12, 36, and 30?

  ○ A. 3   ○ B. 18   ○ C. 12   ○ D. 6

21. What is the least common multiple of 20 and 12?

  ○ A. 4   ○ B. 60   ○ C. 240   ○ D. N

22. Which of the following numbers is composite?

  ○ A. 7   ○ B. 17   ○ C. 51   ○ D. N

23. Which of the following numerals is the same as $\dfrac{25}{40}$?

  ○ A. $\dfrac{1}{2}$ ○ B. $\dfrac{2}{3}$ ○ C. $\dfrac{5}{7}$ ○ D. $\dfrac{5}{8}$

24. Which one of the following decimals is equal to $\dfrac{5}{8}$?

  ○ A. 1.6   ○ B. .625   ○ C. 6.25   ○ D. 16.5

25. If a = 5 and b = 20, what is the value of 2a + b?

  ○ A. 30   ○ B. 40   ○ C. 45   ○ D. 50

26. The product of a number and 6 is 156. What is the number?

  ○ A. 936   ○ B. 162   ○ C. 150   ○ D. 26

27. Look at the following fractions in the box. How many are less than $\dfrac{3}{4}$?

$$\frac{5}{12} , \frac{13}{16} , \frac{2}{3} , \frac{6}{12} , \frac{3}{4} , \frac{3}{8} , \frac{7}{8}$$

  ○ A. 3   ○ B. 4   ○ C. 5   ○ D. 6

28. If x = 12 and y = 10, what is the value of x + x · y?

  ○ A. 120   ○ B. 132   ○ C. 240   ○ D. 1,440

29. $1\dfrac{3}{5} + 6\dfrac{7}{8} =$

  ○ A. $4\dfrac{11}{13}$ ○ B. $8\dfrac{19}{40}$ ○ C. $1\dfrac{23}{40}$ ○ D. $8\dfrac{23}{40}$

30. $2\dfrac{2}{9} - 1\dfrac{7}{18} =$

  ○ A. $3\dfrac{11}{18}$ ○ B. $\dfrac{5}{6}$ ○ C. $1\dfrac{1}{6}$ ○ D. $1\dfrac{5}{6}$

31. $\dfrac{5}{6} \times 7 =$

     ○ A. $\dfrac{35}{36}$ ○ B. $7\dfrac{5}{6}$ ○ C. $5\dfrac{5}{6}$ ○ D. N

32. $\dfrac{3}{4} \times \dfrac{5}{6} \times \dfrac{1}{35} =$

     ○ A. $\dfrac{3}{280}$ ○ B. $1\dfrac{257}{420}$ ○ C. $\dfrac{3}{56}$ ○ D. $\dfrac{1}{56}$

33. $16 \div \dfrac{3}{4} =$

     ○ A. $21\dfrac{1}{3}$ ○ B. $21\dfrac{2}{3}$ ○ C. 12   ○ D. 4

34. $2\dfrac{1}{9} \div 4 =$

     ○ A. 19   ○ B. $\dfrac{23}{36}$ ○ C. $\dfrac{19}{36}$ ○ D. $2\dfrac{1}{9}$

Solve for the variable in each equation:

35. $b + 15 = 24$

     ○ A. 9   ○ B. 11   ○ C. 24   ○ D. 39

36. $8x = 112$

     ○ A. 14   ○ B. 104   ○ C. 120   ○ D. 896

37. $12c = 60$

     ○ A. 192   ○ B. 44   ○ C. 12   ○ D. 5

38. $36 - x = 20$

     ○ A. 6   ○ B. 16   ○ C. 26   ○ D. 56

39. $8x = -32$

     ○ A. −256   ○ B. −40   ○ C. −24   ○ D. −4

40. $\dfrac{a}{0.7} = -2.8$

     ○ A. −4   ○ B. −2.8   ○ C. −1.96   ○ D. 0.7

41. $2y - 4 = 12$

     ○ A. 1   ○ B. 2   ○ C. 4   ○ D. 8

42. What is the area of the rectangle?   4 cm   6.2 cm

     ○ A. 10.2 cm²   ○ B. 24.8 cm²   ○ C. 20.4 cm²   ○ D. N

DIAGNOSTIC TEST

Name _____

43. What is the area of the triangle? 3 cm 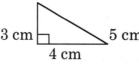 5 cm
    4 cm

  ○ A. 6 cm²   ○ B. 12 cm²   ○ C. 7 cm²   ○ D. 10 cm²

44. What is the area of the circle? (Hint: A = Π r²)

  ○ A. 6.28 in²   ○ B. 12.56 in²   ○ C. 25.12 in²   ○ D. 50.24 in²

45. Which number should come next in this sequence? 1, 3, 7, 13, 21, . . .
   ○ A. 23   ○ B. 29   ○ C. 31   ○ D. 35

46. Prior to landing, an airplane descended 350 feet in 5 minutes. What was the average rate of descent?
   ○ A. 70 ft/min   ○ B. 350 ft/min   ○ C. 1750 ft/min   ○ D. N

47. An auto shop charges $32.50 per hour for repairs. If it takes 2.5 hours to repair your car, what will the charge be for labor?
   ○ A. $13.00   ○ B. $81.25   ○ C. $65.00   ○ D. $812.50

48. A sweater that usually costs $35 is on sale at 25% off. What is the best estimate of the sale price?
   ○ A. $9   ○ B. $26   ○ C. $32   ○ D. $43

49. Chapter 9 of Susan's book starts on page 186 and ends on page 219. How long is the chapter?
   ○ A. 32 pages   ○ B. 33 pages   ○ C. 34 pages   ○ D. 35 pages

50. Dan has 1 brown tie, 2 blue ties, 3 gray ties, and 4 black ties. If he selects a tie without looking, what is the probability that it will be blue?
   ○ A. $\frac{1}{10}$ ○ B. $\frac{1}{5}$ ○ C. $\frac{1}{4}$ ○ D. $\frac{1}{2}$

51. When a six-sided number cube is rolled, what is the probability of rolling a 3 or even number?
   ○ A. $\frac{1}{6}$ ○ B. $\frac{1}{2}$ ○ C. $\frac{2}{3}$ ○ D. $\frac{5}{6}$

52. Stacy's math scores were 76, 84, 73, 78, and 84. What is the mean (average) of these scores?
   ○ A. 77   ○ B. 78   ○ C. 79   ○ D. 84

53. A baseball team had a ratio of 20 wins and 8 losses. What was the ratio of wins to losses?
   ○ A. 2 to 5   ○ B. 5 to 4   ○ C. 5 to 2   ○ D. 5 to 7

POSSIBLE SCORE 53

Computation & Estimation: Estimation Strategies

Estimate each answer.

Strategy 1: Use compatible numbers.

$2,342 \div 8 =$ 2,342 is close to 2,400, and $2,400 \div 8 = 300$; therefore, a good estimate would be **300.**

---

1. $342 \div 7 =$

2. $437 \div 5 =$

3. $532 \div 6 =$

4. $198 \times 24 =$

5. $2,438 \div 53 =$

6. $653 \div 8 =$

7. $724 \div 89 =$

8. $2,779 \div 68 =$

9. $389 \times 12 =$

10. $2,652 \div 88 =$

11. $275 \times 8 =$

12. $1,599 \div 43 =$

Strategy 2: Use rounding. Round numbers to the highest place value.

$33 \times 192 =$ $478 - 319 =$

$30 \times 200 = \textbf{6,000}$ $500 - 300 = \textbf{200}$

13. $59 + 87 =$

14. $92 - 58 =$

15. $865 - 243 =$

16. $149 + 62 =$

17. $807 \times 19 =$

18. $45 \times 39 =$

19. $725 + 487 =$

20. $318 \times 18 =$

21. $12,778 - 7,499 =$

22. $9,704 - 6,218 =$

23. $995 \times 54 =$

24. $873 + 449 =$

Computation & Estimation: Powers & Exponents

$3 \cdot 3 \cdot 3 \cdot 3 \cdot 3 = 3^5$ ↗ *exponent*

↘ *base*

An *exponent* indicates the number of times the *base* is used as a factor.

$5^3 = \underline{5 \cdot 5 \cdot 5}$ Expanded form.
$8 \cdot 8 \cdot 9 \cdot 8 = \underline{8^3 \cdot 9}$ Exponent form.
$6^2 = \underline{36}$ Simplified.

Write each problem in expanded form.

1. 10^5

2. 12^4

3. 2^3

4. 9^6

5. $3^5 \cdot 4^2$

6. $11^2 \cdot 13^3$

7. n^4

8. x^3

9. $y^5 \cdot z^2$

Write in exponent form.

10. $3 \cdot 3 \cdot 3$

11. $12 \cdot 12$

12. $7 \cdot 6 \cdot 7 \cdot 6$

13. $4 \cdot 4 \cdot 2$

14. $8 \cdot 9 \cdot 8 \cdot 9 \cdot 8$

15. $7 \cdot 8 \cdot 7 \cdot 5 \cdot 5$

16. $9 \cdot b \cdot b \cdot a \cdot b$

17. $n \cdot n \cdot m$

18. $x \cdot y \cdot x \cdot y$

Simplify.

19. 10^2

20. $4^3 \cdot 6^2$

21. $5^4 + 2^2$

22. $2^3 \cdot 3^4 \cdot 5^2$

23. $11^2 + 6^3$

24. $6^5 \cdot 9 \cdot 4^2$

Equations & Inequalities: Writing Algebraic Expressions & Equations

Part I: Write an algebraic expression for each phrase or problem.

7 more points scored than the Panthers

*Let **p** represent the number of **p**oints scored by the Panthers.*
*The words **more than** suggest addition.*

p + 7

TIP: *Be sure to identify whether you should add, subtract, multiply or divide.*

---

1. the sum of 8 and y

2. 6 more than m

3. nine less than r

4. the difference of 6 and x

5. n divided by 9

6. 5 taken away from y

7. the quotient of c and 4

8. the product of a number and 12

9. 8 points less than Tom's score

10. the sum of 11 and x

Part II: Translate each problem into an algebraic equation.

A number minus 6 equals 12.   Let n = number. $n - 6 = 12$

TIP: *Remember that there is an equals sign in every equation.*

11. A number minus 8 equals 10.

12. 6 more than 9 times the number of fish is 24.

13. The product of 6 and x added to 12 is 24.

14. The sum of 7 and the quotient of 10 and a number is 12.

15. The cost of a hamburger plus 7 cents tax is $2.09.

16. 8 less than the product of 4 and y is 32.

Equations & Inequalities: Evaluating Expressions

Evaluate each algebraic expression.

$c \div d - 4$ for $c = 15$ and $d = 3$
$15 \div 3 - 4 = 1$

1. Substitute the numbers for variables.
2. Evaluate.

Evaluate each expression for $x = 5$, $y = 8$, and $z = 9$.

1. $(y + y) - 8$

2. $z - (13 - y)$

3. $z - x$

4. $x + y^2$

5. $z - x + y$

6. $xy - 10$

7. $(x + y) - (y - x)^2$

8. $40 \div y$

9. $3z - 2y$

Evaluate each expression for $a = 5$, $b = 10$, $c = 2$, and $d = 3$.

10. $\dfrac{ab}{c}$

11. $bc^2 - d$

12. $\dfrac{4d}{2}$

13. $(b - a) + (10 - c)$

14. $a^2 - d$

15. abc^2

16. $(bc)^2 \div a$

17. $b - cd$

18. $(ab)(cd)$

 THINK ABOUT IT!

19. Find number replacements for a and b so that $a + b = 16$ and $a - b = 2$.

Equations & Inequalities: Solving Addition & Subtraction Equations

Solve and check each equation.

$$x + {}^-15 = 30$$
$$x + {}^-15 - {}^-\underline{15} = 30 - {}^-\underline{15}$$
$$x = 45$$
Check: $45 + {}^-15 = 30$
$$30 = 30 ✔$$

1. Look at what has been done to the variable.
2. Undo it using the inverse (opposite) operation on both sides of the equation.
3. To check, replace the variable with your solution.

TIP: *When negative numbers are involved, do not change the sign of the integer. Just add or subtract the negative number.*

---

1. $c - 40 = 10$

2. $y + 28 = 51$

3. $16 + x = 49$

4. $70 + r = 82$

5. $x - 22 = 61$

6. $n + 81 = 460$

7. $c - 86 = 104$

8. $n - 238 = 376$

9. $547 = n + 82$

10. $^-74 = n + 17$

11. $^-263 = n - 35$

12. $s - 76 = 275$

13. $x + 267 = 194$

14. $67 + x = {}^-382$

15. $^-65 + x = {}^-100$

16. $n + (^-37) = 94$

17. $x - 46 = {}^-298$

18. $m - (^-75) = 60$

19. $x + 89 = 67$

20. $^-9 = x - 36$

21. $27 = c + 91$

22. $36 + r = 11$

23. $x + 67 = 204$

24. $r - 12 = {}^-31$

Equations & Inequalities: Solving Multiplication & Division Equations

Solve and check each equation.

$-15n = 270$
$-15n \div -15 = 270 \div -15$
$n = -18$
Check: $-15 \times -18 = 270$
$270 = 270$ ✔

1. Look at what has been done to the variable.
2. Undo it using the inverse (opposite) operation on both sides of the equation.
3. To check, replace the variable with your solution.

TIP: *Remember when multiplying or dividing integers, if the signs are the same, the answer is positive. If the signs are different, the answer is negative.*

---

1. $16x = 288$

2. $207 = 9x$

3. $12h = 312$

4. $n \div 42 = 15$

5. $36x = 288$

6. $n \div 23 = 18$

7. $\dfrac{n}{19} = 13$

8. $\dfrac{n}{64} = 11$

9. $\dfrac{n}{12} = -104$

10. $12y = 252$

11. $-930 = 15x$

12. $-26y = 910$

13. $n \div -31 = 14$

14. $-368 = 16c$

15. $-105 = -7y$

16. $\dfrac{n}{-9} = -387$

17. $\dfrac{h}{85} = -125$

18. $\dfrac{x}{-26} = 7$

19. $n \div 17 = 9$

20. $24x = -144$

21. $-272 = -34s$

22. $r \div 18 = -15$

23. $-31c = 372$

24. $n \div -21 = -19$

Equations & Inequalities: Solving Two-Step Equations

Solve and check the following equations.

$8n - 21 = 75$

$8n - 21 \underline{+ 21} = 75 \underline{+ 21}$
$8n = 96$
$8n \underline{\div 8} = 96 \div 8$
$n = 12$

When solving a two-step equation, undo each operation.
1. First, identify the order in which the operations have been applied to the variable.
2. Then, undo each operation in reverse order.

$\underline{8 \cdot 12} - 21 = 75$
$96 - 21 = 75$
$75 = 75$ ✔

To check, replace the variable with the solution.

---

1. $6n + 36 = 144$

2. $\dfrac{x}{4} + 8 = 18$

3. $4y + 16 = 28$

4. $4x - 45 = 39$

5. $7n - 13 = 50$

6. $\dfrac{x+8}{9} = {}^-6$

7. $12x - 23 = 37$

8. $\dfrac{x}{-3} + 13 = {}^-4$

9. $4r - 8 = 32$

10. $\dfrac{h}{12} + 7 = 28$

11. $\dfrac{n}{-6} - 17 = {}^-8$

12. $\dfrac{n-6}{-5} = 3$

13. $5n + 3 = {}^-22$

14. $4r - 7 = {}^-15$

15. $5e - 8 = 42$

16. $\dfrac{x+8}{6} = 7$

17. $\dfrac{x-7}{12} = {}^-16$

18. $\dfrac{d-5}{7} = 14$

 THINK ABOUT IT!

19. Solve each equation.

a. $\dfrac{3c-5}{2} = 8$

b. $\dfrac{12y-12}{8} = {}^-9$

c. $\dfrac{3x-4}{4} = 2$

Equations & Inequalities: Solving Inequalities

Solve and check the following inequalities.

To solve an inequality, use the same procedure as when solving equations. *However, when you **multiply** or **divide** by a negative number, reverse the sign of the inequality.*

$$x - 7 \geq {}^-8$$
$$x - 7 + 7 \geq {}^-8 + 7$$
$$x \geq {}^-1$$

$$^-3x < 12$$
$$^-3x \div {}^-3 < 12 \div {}^-3$$
$$x > {}^-4$$

---

1. $4 < x - 5$

2. $c - 1 < 4$

3. $x - 4 > 8$

4. $^-4m > -2$

5. $\dfrac{c}{5} < 15$

6. $16 > \dfrac{n}{4}$

7. $^-7x < 35$

8. $^-25 < 5n$

9. $^-2x > 10$

10. $x - 8 > 0$

11. $^-9 + x < 10$

12. $^-9x \geq 108$

13. $4c + 2 > -22$

14. $10 \geq 5(y - 1)$

15. $3a - 1 > 5$

16. $^-3x + 24 > -3$

17. $2(n - 6) \geq 4$

18. $2a + 3 \geq 12$

19. $6 + 2n < {}^-1$

20. $3(x + 1) < 12$

21. $2y + 4 < 16$

 THINK ABOUT IT!

22. The product of an integer and negative three is greater than $^-21$. Find the greatest integer that satisfies this condition.

Equations & Inequalities: Problem Solving   Name _____

Solve each problem.

1. Together two items cost $21. One item   1. _____
   costs $6. What does the other item cost?

2. Stacy worked 36 hours last week and   2. _____
   earned $234. How much did she earn per
   hour?

3. If H represents Dr. Harrison's age, and S   3. _____
   represents Dr. Schultz's age, explain
   what the sentence H = S + 3 means.

4. Write and solve an equation in which the   4. _____
   product is 522; one factor is 3, and the
   other factor is represented by x.

5. Write and solve an equation in which y is   5. _____
   divided by 8 and the quotient is 5.

6. If y > 4 and y < 12, explain what 4 < y < 12 6. _____
   means.

7. Find at least two values for **a** that make   7. _____
   2a + 3 ≥ 7 true. Then find two values that
   make it false.

8. Describe the solution of each equation.   8. _____
   a. x = x
   b. y + 4 = 3
   c. c + 1 = c + 2

Decimals: Comparing & Ordering

Name _____

Part I: Compare. Use >, <, or = for each

7.001 ◯ 7.010

To compare 2 decimals:
1. Line up the decimal points.
2. Compare digits from *left to right* in their corresponding place value.

7.0⓪1
7.0①0

0 < 1 so 7.001 < 7.010

TIP: *When comparing negative numbers, remember that the number closest to zero has the highest value.*

---

1. 47.14 ◯ 47.13

2. ⁻7.003 ◯ ⁻7.03

3. 18.06 ◯ 18.060

4. 1.01 ◯ 1.101

5. ⁻3.658 ◯ ⁻3.685

6. ⁻15.3 ◯ ⁻15.29

Part II: Order from least to greatest.

0.004, 0.039, 0.0041

0.004
0.039
0.0041
Answer: 0.004, 0.0041, 0.039

To compare 2 or more decimals:
1. Line up the decimal points.
2. Starting from the left, compare the digits in their corresponding place value.

TIP: *Comparing two at a time to find the least or greatest may help.*

7.   3.7, 3.07, 3.069

8.   4.01, 4, 4.001

9.   ⁻6.404, ⁻6.044, ⁻6.04

10.   ⁻0.001, ⁻0.101, ⁻0.011

11.   20.03, 20.303, 20.003

12.   ⁻12.11, ⁻12.01, ⁻12.10

Decimals: Estimating

Name _____

Estimate by rounding.

Round to the nearest . . .

whole number	*tenth*	*hundredth*
19.501 = 20	42.73 = 42.7	6.493 = 6.49

Estimate by rounding to the nearest whole number.

1. $31.53 + 17.4$

2. $21.54 - 18.91$

3. 19.8×4.7

4. $71.49 - 3.8$

5. $44.5 + 16.39$

6. $17.8 \div 2.13$

7. 5.89×7.15

8. $32.38 \div 4.21$

9. $16.8 + 22.3$

Estimate by rounding to the nearest tenth.

10. $14.35 - 6.13$

11. $21.051 - 13.034$

12. $16.895 + 18.333$

13. $32.875 - 19.076$

14. $7.397 + 10.143$

15. $81.005 + 16.751$

16. $14.603 - 12.2571$

17. $31.313 + 14.056$

18. $41.358 - 17.159$

Estimate by rounding to the nearest hundredth.

19. $8.9945 + 17.6513$

20. $3.0054 - 2.1067$

21. $16.8901 + 9.6767$

22. $21.8945 + 19.01011$

23. $7.91912 - 6.81012$

24. $81.8395 - 70.4227$

Decimals: Adding & Subtracting

Name_____

Find each sum or difference. *Watch the sign!*

24.6 − 5.73

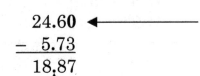

$$
\begin{array}{r}
24.60 \\
-\ 5.73 \\
\hline
18.87
\end{array}
$$

1. Line up the decimal points
2. Add zeros if necessary.
3. Add or subtract as with whole numbers.

TIP: *Remember to bring your decimal point down into your answer.*

---

1. $\begin{array}{r} 35.82 \\ -\ 12.3\ \ \\ \hline \end{array}$

2. $\begin{array}{r} 3.573 \\ +\ 7.41\ \ \\ \hline \end{array}$

3. $\begin{array}{r} 515 \\ -\ 408.71 \\ \hline \end{array}$

4. $\begin{array}{r} 45.001 \\ -\ 36.987 \\ \hline \end{array}$

5. $\begin{array}{r} 202.554 \\ +\ 39.75 \\ \hline \end{array}$

6. $\begin{array}{r} 62.1 \\ -\ 0.0034 \\ \hline \end{array}$

7. $\begin{array}{r} 3.056 \\ -\ 2.678 \\ \hline \end{array}$

8. $\begin{array}{r} 227 \\ -\ 205.78 \\ \hline \end{array}$

9. $\begin{array}{r} 0.09801 \\ +\ 3.78502 \\ \hline \end{array}$

10. 7.9 + (⁻3.8)

11. 9.73 + (⁻4.7)

12. ⁻21.4 − 15.72

13. ⁻4.9 + (⁻3.2)

14. ⁻2.3 + 5.7

15. 79.34 − (⁻32.48)

16. 4.5 + (⁻4.5)

17. ⁻10.28 − (⁻5.8)

18. 6.73 + (⁻5.82)

19. ⁻28.15 + 7.24

20. ⁻2.1 + ⁻3.6

21. 6.7 + ⁻9.1

Decimals: Multiplying & Dividing

Find each product.

19.5 × .19

19.5	*1 decimal place*
× .19	*+ 2 decimal places*
1755	
1950	
3.705	*3 decimal places*

1. Multiply as you would whole numbers.
2. The number of decimal places in the product is the sum of the decimal places in the factors.

TIP: *Remember, do not line up the decimal points when setting up your problem.*

---

1. 8.1
   × 4.5

2. 3.9
   × .07

3. 16.32
   × 1.03

4. 3.46
   × 5.7

5. ⁻0.32 × 4.01

6. ⁻0.01 × 5280

7. ⁻84.6 × (⁻3.05)

8. ⁻75.2 × 100

Find each quotient.
3.9 ÷ .13

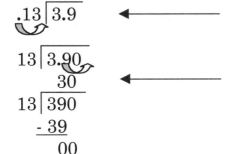

1. Change the divisor to a whole number by moving the decimal point to the right.
2. Move the decimal point in the dividend the same number of spaces. Add zeros if necessary.
3. Divide.

---

9. 1.3 ⟌ 4.42

10. 12.8 ⟌ 46.08

11. 2.4 ⟌ 27.12

12. 0.18 ⟌ 2.88

13. 10.4 ÷ (⁻0.02)

14. ⁻51.3 ÷ (⁻1.9)

15. 1.69 ÷ (⁻1.3)

16. ⁻3.75 ÷ 0.25

29

Decimals: Solving Equations

Name _____

Solve and check each equation. Remember to use the inverse operation to solve the variable.

$$w + 0.05 = 7.95$$
$$w + 0.05 \underline{- 0.05} = 7.95 \underline{- 0.05} \qquad \textit{Subtract 0.05 from both sides.}$$
$$w = 7.9$$

---

1. $38.06 + x = 81.5$

2. $m - 17.9 = 483.6$

3. $36.7 = y + 18.4$

4. $1.005x = 20.1$

5. $1.41x = {}^-9.87$

6. $23.5a = 9.4$

7. $48 = \dfrac{a}{3.2}$

8. $\dfrac{x}{8.3} = {}^-2.3$

9. $\dfrac{c}{6.5} = 7.2$

10. $3.2 = \dfrac{y}{5.4}$

11. $^-120 = \dfrac{h}{2}$

12. $^-14.95 = t + 25.87$

13. $29.6 + x = 142.7$

14. $x - 39.4 = {}^-46.7$

15. $4.5m = {}^-67.5$

16. $x + 1.5 = 5.03$

17. $85 + p = {}^-200$

18. $0 = {}^-5r$

19. $x + 3.5 = {}^-12.1$

20. $^-0.15c = 0.255$

21. $^-81 = .9t$

 THINK ABOUT IT!

$$x = \frac{-b \pm \sqrt{b^2 - 4ac}}{2a}$$

22. Use inverse operations to find the missing number.

Begin with x ➤ multiply by 1.4 ➤ add 5.2 ➤ end with 14.16.

30

Decimals: The Metric System

Name _____

Study the chart below, then complete each sentence.

$\times 10$	$\times 10$	$\times 10$	$\times 10$	$\times 10$	$\times 10$

meter

kilo-   hecto-   deka-   *liter*   deci-   centi-   milli-

$\div 10$ $\div 10$ $\div 10$ *gram* $\div 10$ $\div 10$ $\div 10$

$$6.3 \text{ m} = \underline{630} \text{ cm} \qquad\qquad 25 \text{ cm} = \underline{.00025} \text{ km}$$
$$6.3 \times 10^2 = 630 \qquad\qquad\qquad 25 \div 10^5 = .00025$$

TIP: *When multiplying by a power of ten, simply move the decimal point to the right; when dividing, move it to the left.*

---

1. $4.2 \text{ m} = $ _____ cm

2. $4000 \text{ mg} = $ _____ g

3. $6.7 \text{ L} = $ _____ mL

4. $9.14 \text{ hm} = $ _____ m

5. $8 \text{ cm} = $ _____ m

6. $3 \text{ m} = $ _____ mm

7. $420 \text{ mL} = $ _____ L

8. $0.73 \text{ m} = $ _____ cm

9. $500 \text{ cm} = $ _____ m

10. $4.02 \text{ g} = $ _____ cg

11. $3{,}183 \text{ mm} = $ _____ m

12. $13.2 \text{ m} = $ _____ cm

13. $6 \text{ km} = $ _____ m

14. $5.4 \text{ g} = $ _____ mg

15. $0.41 \text{ kg} = $ _____ mg

16. $7300 \text{ L} = $ _____ kL

17. $423 \text{ m} = $ _____ km

18. $320 \text{ cm} = $ _____ km

 THINK ABOUT IT!

19. Write a general rule for converting in the metric system.

Decimals: Scientific Notation

Name _____

Part I: Write each number in scientific notation.

8,250,000.

$\underline{8.25 \times 10^6}$

1. Move the decimal point to change your number (n) so that $1 \le n < 10$.

0.000064

$\underline{6.4 \times 10^{-5}}$

2. Count how many places you moved the decimal point. Place that number as your power of 10 (*positive* if you moved to the left, *negative* if you moved to the right).

---

1.   340

2.   0.0068

3.   5,400

4.   7,180,000

5.   0.00002

6.   5,600,000

7.   97,000

8.   0.00084

9.   23,000

Part II. Write each number in standard form.

5.2×10^5
520,000

7.8×10^{-4}
.00078

1. Look at the exponent on the 10.
2. If it is *positive*, move the decimal point that many places to the right.
3. If it is *negative*, move it that many places to the left.

---

10. 3.2×10^5

11. 5.203×10^3

12. 2.43×10^{-3}

13. 8×10^8

14. 3.19×10^{-6}

15. 2.83×10^{-4}

16. 7.4×10^{-5}

17. 29.8×10^2

18. 6.7×10

 THINK ABOUT IT!

19. Place the following in order from the smallest to the largest.

5.1×10^{-5} 5.12×10^{-5} 5.1×10^{-6}

Decimals: Problem Solving

Solve each problem.

1. Robbie already has $376.25 in his
   checking account, then he receives his
   paycheck for $135.90. He has two bills to
   pay; one for $297.28 and the other for
   $325.30. DoesRobbie have enough money
   in his account to pay both bills? Explain.

1.

2. Use estimation to decide which is a better
   buy: 5 candy bars for $1.98 or 3 candy
   bars for $1.10.

2.

3. Harry drove 327 miles on 12.9 gallons of
   gasoline. To the nearest tenth, how many
   miles per gallon did Harry's car get?

3.

4. Suppose the directions for cooking a
   turkey are 30 minutes per pound. If you
   have a 9.7 lb turkey, to the nearest
   minute, how long should you cook the
   turkey?

4.

5. Which has a greater capacity, a 2-liter
   bottle of soda or a six-pack of cans each
   containing 354 mL?

5.

6. How many seconds are there in a 365-day
   year? Express your answer in scientific
   notation.

6.

7. For her science class, Mrs. Kihl bought 24
   pairs of safety goggles at $2.95 each, 6
   beakers at $1.95 each, and 2 dozen test
   tubes at $6.95 per dozen. How much
   money did she spend in all?

7.

Number Theory: Squares & Square Roots   Name _____

Part I: Find the square of each number.

When you compute 5 x 5 or 5^2, you are finding the square of 5.

---

1. 3	2. 8	3. 15	4. 13	5. 9
6. 7	7. 14	8. 20	9. 10	10. 16
11. 11	12. 17	13. 25	14. 18	15. 12

Part II: Using a calculator, find each square root. Round answers to the nearest thousandth if necessary.

*The symbol that is used to find a nonnegative square root is $\sqrt{}$, which is called a **radical sign**.*

$\sqrt{49} = 7$ Since $7^2 = 49$, 7 is the **square root** of 49.

---

16. $\sqrt{25}$	17. $\sqrt{64}$	18. $\sqrt{81}$	19. $\sqrt{121}$	20. $\sqrt{9}$
21. $\sqrt{100}$	22. $\sqrt{421}$	23. $\sqrt{92}$	24. $\sqrt{848}$	25. $\sqrt{144}$
26. $\sqrt{400}$	27. $\sqrt{625}$	28. $\sqrt{87}$	29. $\sqrt{57}$	30. $\sqrt{125}$

 **THINK ABOUT IT!**

31. A square has an area of 8 cm². To the nearest tenth, what is the length of a side of the square?

Number Theory: Multiples & Factors   Name _____

Part I: Write the first four nonzero multiples of each number.

> *A multiple of a number is the product of two whole numbers.*
>
> The first four nonzero multiples of 5 are
> $5 \times 1 = \mathbf{5}$, $5 \times 2 = \mathbf{10}$, $5 \times 3 = \mathbf{15}$, and $5 \times 4 = \mathbf{20}$.

1. 3	2. 7	3. 2	4. 8
5. 6	6. 9	7. 10	8. 20
9. 11	10. 25	11. 14	12. 15

Part II: Find all the factors of each number.

> *A factor is a number that when multiplied by another number, gives a product.*
>
> $20 = 1 \times 20$, 2×10, 4×5
> factors = 1, 2, 4, 5, 10, 20

13. 16	14. 28	15. 84	16. 40
17. 75	18. 34	19. 60	20. 105

 THINK ABOUT IT!

21. Solve each equation. Use the results to list all the factors of 24.

         a. $1m = 24$ c. $3x = 24$

         b. $2y = 24$ d. $4c = 24$

Number Theory: Prime & Composite Numbers

Tell whether the following is prime (P) or composite (C). If the number is composite give at least one factor of the composite number other than 1 and itself.

A **prime** number is a number greater than one that has exactly two factors, 1 and itself. $5 = 1 \times 5$ $11 = 1 \times 11$

A **composite** number is a number greater than one that has more than two factors. $9 = 1 \times 9$ and 3×3 (3 factors)

---

1. 22

2. 14

3. 18

4. 19

5. 23

6. 27

7. 37

8. 15

9. 12

10. 16

11. 25

12. 29

13. 40

14. 77

15. 51

16. 17

17. 45

18. 47

19. 100

20. 99

21. 63

22. 70

23. 38

24. 44

25. 73

26. 13

27. 54

28. 26

 THINK ABOUT IT!

29. Find the smallest number for x that will give a composite value for 6x – 1.

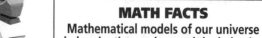

MATH FACTS
Mathematical models of our universe help scientists understand the behavior of stars, planets and galaxies. Computers generate pictures of our universe by processing data, gathered from giant telescopes, with mathematical equations.

Number Theory: Prime Factorization

Name _____

Use a factor tree to find the prime factorization of each number. Use exponents when possible.

Prime factorization of a composite number is taking that number and expressing it as a product of all prime factors. (Prime numbers have only two factors, 1 and the number itself, and composite numbers have more than two factors.)

Factor tree:

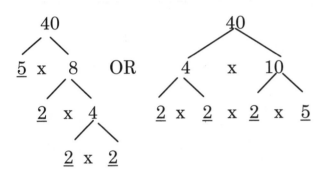

Prime Factorization

$2 \cdot 2 \cdot 2 \cdot 5 = 2^3 \cdot 5$

---

1. 36

2. 90

3. 40

4. 85

5. 51

6. 225

 THINK ABOUT IT!

7. Solve each equation to find the prime factorization of each number.

   a. $80 = 2^3 \cdot 5 \cdot n$ b. $72 = 2^2 \cdot 3^2 \cdot x$ c. $2^2 \cdot 5^2 \cdot n = 300$

Find the Greatest Common Factor (GCF) of each pair of numbers.

12, 20   1. List all factors of each number.

12: 1, 2, 3, ④, 6, 12   2. Find the greatest factor common to both
20: 1, 2, ④ 5, 10 , 20   numbers.
           GCF = 4

---

1.   16, 24   2.   10, 21   3.   32, 40

4.   84, 108   5.   9, 10   6.   36, 54

7.   30, 45   8.   120, 182   9.   30, 105

10.   30, 36, 48   11.   16, 28, 40   12.   12, 18, 24

 **THINK ABOUT IT!**

13. Find the GCF of each pair of expressions.
     a. $9x^2y^3$ and $12xy^2$ b. $2y^2z^3$ and $10x^2z^2$ c. $8xy^2z^3$ and $11x^2yz$

Number Theory: Lowest Common Multiple   Name _____

Find the Lowest Common Multiple (LCM) for each pair of numbers.

9, 15

9: 9, 18, 27, 36, (45), 54
15: 15, 30, (45)
LCM = 45

List multiples in order for each number until a multiple is found that is common to both numbers. This will be the LCM.

---

1. 8, 9

2. 5, 7

3. 42, 70

4. 20, 24

5. 60, 80

6. 18, 300

7. 10, 15

8. 30, 625

9. 8, 20

10. 11, 22, 33

11. 14, 21, 84

12. 9, 12, 15

13. 6y, 10y

14. 8x, 24y

15. 4xy, 10y

Number Theory: Problem Solving

Name _____

Solve each problem.

1. Describe three ways in which 18 desks can be arranged in rows in a classroom with the same number of desks in each row, with more than 1 row and more than 2 desks in each row.

2. Consecutive primes, such as 5 and 7, that have a difference of 2, are called twin primes. Find five more pairs of twin primes.

3. Using your calculator, try to find the square root of $^-36$. What do you notice on the display? Why do you suppose this happens?

4. The area of a square is 12 cm². To the nearest tenth, estimate the length of a side of the square.

5. List all of the factors of 20 and 30. Then, list all of the factors of the least common multiple of 20 and 30. What do you notice?

6. Two numbers are considered to be relatively prime if their GCF is 1. Are the following sets of numbers relatively prime? If not, explain why.
   a. 8 and 9
   b. 20 and 25
   c. 35 and 56

7. Find the smallest value of x that gives a composite value for:
   a. 12x − 1
   b. 12x + 1

1.
2.
3.
4.
5.
6.
7.

Rational Numbers: Simplifying Fractions   Name _____

Express each fraction in simplest form.

$\dfrac{24}{36}$ To write a fraction in simplest form, divide its numerator and denominator by their GCF and write the resulting fraction. $\dfrac{24 \div 12}{36 \div 12} = \dfrac{2}{3}$

TIP: *Remember: A fraction is in lowest terms when the GCF of its numerator and denominator is 1.*

___

1. $\dfrac{40}{52}$ 2. $\dfrac{66}{102}$ 3. $\dfrac{6}{54}$ 4. $\dfrac{45}{72}$

5. $\dfrac{8}{9}$ 6. $\dfrac{30}{40}$ 7. $\dfrac{14}{28}$ 8. $\dfrac{56}{80}$

9. $\dfrac{27}{33}$ 10. $\dfrac{42}{90}$ 11. $\dfrac{75}{90}$ 12. $\dfrac{63}{81}$

13. $\dfrac{24}{40}$ 14. $\dfrac{9}{51}$ 15. $\dfrac{30}{35}$ 16. $\dfrac{12}{40}$

17. $\dfrac{21}{45}$ 18. $\dfrac{110}{156}$ 19. $\dfrac{84}{128}$ 20. $\dfrac{49}{140}$

21. $\dfrac{68}{76}$ 22. $\dfrac{180}{270}$ 23. $\dfrac{71}{82}$ 24. $\dfrac{77}{121}$

 THINK ABOUT IT!

25. Fractions can be reduced with variables the same way they can be reduced with numbers. Reduce the following fractions.

a. $\dfrac{3x}{4x}$ b. $\dfrac{xyz}{2xz}$ c. $\dfrac{13xy}{26x}$

Rational Numbers: Improper Fractions & Mixed Numbers

Part I: Change each improper fraction to a mixed number in simplest form or a whole number.

$\dfrac{19}{4}$

1. Divide the numerator by the denominator.

$\begin{array}{r} 4 \\ 4\overline{)19} \\ -16 \\ \hline 3 \quad 4\frac{3}{4} \end{array}$

2. If there is a remainder, put it in fraction form over the divisor.

3. Reduce fraction to lowest terms.

1. $\dfrac{81}{12}$

2. $\dfrac{27}{11}$

3. $\dfrac{45}{10}$

4. $\dfrac{74}{8}$

5. $\dfrac{51}{3}$

6. $\dfrac{24}{4}$

7. $\dfrac{36}{7}$

8. $\dfrac{7}{4}$

9. $\dfrac{8}{8}$

10. $\dfrac{70}{16}$

11. $\dfrac{390}{25}$

12. $\dfrac{6}{2}$

13. $\dfrac{22}{5}$

14. $\dfrac{23}{7}$

15. $\dfrac{215}{15}$

Part II: Change each mixed number or whole number to an improper fraction.

$8\frac{1}{2}$

$8 \times 2 + 1 = \dfrac{17}{2}$

1. Multiply the whole number by the denominator.

2. Add the numerator.

3. Place that number over the denominator.

16. $5\frac{7}{8}$

17. $8\frac{2}{3}$

18. $8\frac{7}{11}$

19. $3\frac{6}{15}$

20. $3\frac{7}{8}$

21. $4\frac{31}{40}$

22. $3\frac{4}{5}$

23. $2\frac{3}{8}$

24. $6\frac{3}{5}$

25. $4\frac{3}{51}$

26. $11\frac{2}{11}$

27. $6\frac{3}{8}$

28. $6\frac{2}{7}$

29. $8\frac{4}{7}$

30. $8\frac{9}{10}$

Rational Numbers : Comparing and Ordering   Name _____

Part 1:   Write <, >, or = for each $\bigcirc$.

$\dfrac{3}{5}$ $\bigcirc$ $\dfrac{4}{6}$ To compare two fractions, find their cross products and compare.

   18<20, so $\dfrac{3}{5}$ $\underset{<}{\bigcirc}$ $\dfrac{4}{6}$

---

1. $\dfrac{3}{10}$ $\bigcirc$ $\dfrac{1}{3}$

2. $\dfrac{17}{30}$ $\bigcirc$ $\dfrac{3}{5}$

3. $\dfrac{2}{5}$ $\bigcirc$ $\dfrac{3}{7}$

4. $\dfrac{5}{13}$ $\bigcirc$ $\dfrac{6}{15}$

5. $\dfrac{2}{3}$ $\bigcirc$ $\dfrac{8}{11}$

6. $\dfrac{-5}{6}$ $\bigcirc$ $\dfrac{-6}{8}$

7. $\dfrac{-5}{8}$ $\bigcirc$ $\dfrac{-7}{10}$

8. $\dfrac{-4}{5}$ $\bigcirc$ $\dfrac{-5}{7}$

9. $\dfrac{18}{24}$ $\bigcirc$ $\dfrac{10}{18}$

10. $\dfrac{17}{20}$ $\bigcirc$ $\dfrac{9}{11}$

11. $\dfrac{1}{3}$ $\bigcirc$ $\dfrac{2}{4}$

12. $\dfrac{-5}{12}$ $\bigcirc$ $\dfrac{-3}{7}$

13. $\dfrac{14}{15}$ $\bigcirc$ $\dfrac{16}{17}$

14. $\dfrac{13}{12}$ $\bigcirc$ $\dfrac{8}{7}$

15. $\dfrac{-5}{8}$ $\bigcirc$ $\dfrac{-4}{9}$

Part II: Order from least to greatest.

TIP: *Try changing the fractions to decimals, then compare them.*

16. $\dfrac{-7}{8}$, $\dfrac{-5}{7}$, $\dfrac{-3}{4}$

17. $\dfrac{2}{3}$, $\dfrac{2}{7}$, $\dfrac{2}{5}$, $\dfrac{2}{9}$

18. $\dfrac{5}{6}$, $\dfrac{6}{7}$, $\dfrac{3}{8}$, $\dfrac{3}{7}$

19. $\dfrac{-5}{12}$, $\dfrac{-6}{15}$, $\dfrac{-3}{4}$, $\dfrac{-2}{3}$

20. $\dfrac{9}{10}$, $\dfrac{4}{5}$, $\dfrac{8}{9}$, $\dfrac{5}{6}$

21. $\dfrac{1}{5}$, $\dfrac{1}{4}$, $\dfrac{1}{3}$

Rational Numbers: Adding and Subtracting

Find each sum or difference. Reduce.

$$\frac{-5}{6} = \frac{-20}{24}$$
$$+ \frac{3}{8} = \frac{9}{24}$$
$$\frac{-11}{24}$$

1. Find the lowest common denominator (LCD).

2. Write the equivalent fractions using the LCD.

3. Add or subtract the numerators. Write the sum or difference over the LCD. Reduce if necessary.

TIP: *Pay close attention when adding or subtracting negative fractions.*

---

1. $\dfrac{1}{6}$
    $+ \dfrac{3}{4}$

2. $\dfrac{-8}{16}$
    $+ \dfrac{3}{4}$

3. $\dfrac{2}{3}$
    $- \dfrac{1}{12}$

4. $\dfrac{3}{4}$
    $+ \dfrac{5}{6}$

5. $\dfrac{-2}{5}$
    $- \dfrac{3}{4}$

6. $\dfrac{3}{4}$
    $- \dfrac{4}{7}$

7. $\dfrac{5}{8}$
    $+ \dfrac{5}{12}$

8. $\dfrac{-7}{10}$
    $- \dfrac{1}{2}$

9. $\dfrac{-11}{12} - \dfrac{-5}{3} =$

10. $\dfrac{8}{14} - \dfrac{-4}{6} =$

11. $\dfrac{-1}{8} - \dfrac{-5}{12} =$

12. $\dfrac{-2}{3} + \dfrac{7}{9} =$

13. $\dfrac{5}{6} + \dfrac{7}{10} =$

14. $\dfrac{4}{5} + \dfrac{-3}{8} =$

15. $\dfrac{7}{8} - \dfrac{-5}{6} =$

16. $\dfrac{-8}{12} - \dfrac{2}{3} =$

Rational Numbers: Adding and Subtracting   Name _____

Find each sum or difference. Reduce.

$$-4\frac{1}{6} = \frac{-25}{6} = \frac{-50}{12}$$
$$+3\frac{1}{4} = \frac{13}{4} = \frac{39}{12}$$
$$\frac{-11}{12}$$

1. Change to improper fractions.
2. Find the lowest common denominator (LCD).
3. Add or subtract. Reduce if necessary.

---

1. $1\frac{1}{2}$
   $+ \quad \frac{3}{4}$

2. $2\frac{3}{8}$
   $- 1\frac{7}{8}$

3. $-6\frac{1}{4}$
   $+ 4\frac{1}{2}$

4. $5\frac{5}{12}$
   $- 3\frac{2}{3}$

5. $3\frac{1}{6}$
   $+ \quad \frac{5}{12}$

6. $4\frac{3}{10}$
   $- 3\frac{1}{2}$

7. $\frac{9}{10}$
   $+ 2\frac{1}{3}$

8. $-7\frac{1}{2}$
   $- 3\frac{1}{4}$

9. $12\frac{1}{5}$
   $+ 11\frac{7}{8}$

10. $6\frac{1}{4}$
    $- 3\frac{7}{10}$

11. $4\frac{7}{12}$
    $+ 6\frac{4}{9}$

12. $4\frac{2}{9}$
    $- 1\frac{1}{6}$

13. $3\frac{5}{9} - \left(-2\frac{3}{5} \right) =$

14. $3\frac{7}{10} + \left(-3\frac{7}{10} \right) =$

15. $12\frac{5}{8} - \left(-3\frac{2}{5} \right) =$

16. $2\frac{1}{2} + \left(-3\frac{2}{7} \right) =$

17. $3\frac{5}{9} - \left(-2\frac{3}{5} \right) =$

18. $-3\frac{1}{3} - \left(-4\frac{2}{5} \right) =$

Rational Numbers: Solving Equations   Name _____

Solve and check each equation.

$$n - \frac{6}{8} = \frac{2}{3}$$

$$n - \frac{6}{8} + \frac{6}{8} = \frac{2}{3} + \frac{6}{8}$$

$$n = 1\frac{5}{12}$$

1. Look at what has been done to the variable.

2. Undo it by using the inverse (opposite) operation on both sides of the equation.

$$1\frac{5}{12} - \frac{6}{8} = \frac{2}{3}$$

$$\frac{17}{12} - \frac{6}{8} = \frac{2}{3}$$

$$\frac{34}{24} - \frac{18}{24} = \frac{16}{24} = \frac{2}{3} \checkmark$$

3. Check your answer by plugging it back into the equation to see if it makes the equation true.

---

1. $x - \frac{7}{2} = \frac{1}{2}$

2. $x + \frac{1}{5} = \frac{-3}{4}$

3. $m - \frac{3}{5} = \frac{5}{12}$

4. $n - \frac{7}{8} = 3\frac{1}{4}$

5. $x - 3\frac{1}{2} = -3\frac{1}{2}$

6. $c + \frac{12}{25} = \frac{3}{10}$

7. $y - \frac{3}{5} = -\frac{1}{10}$

8. $x + \frac{2}{3} = \frac{5}{4}$

9. $-\frac{9}{10} = \frac{5}{12} + y$

10. $-\frac{4}{5} = \frac{5}{6} + x$

11. $n - \frac{1}{10} = \frac{3}{20}$

12. $\frac{3}{7} = x + \frac{2}{5}$

13. $\frac{-1}{5} + y = \frac{1}{4}$

14. $\frac{-4}{9} = w + \frac{5}{8}$

15. $x - \left(\frac{-1}{7} \right) = \frac{5}{14}$

46

Rational Numbers: Multiplying Fractions and Mixed Numbers

Find each product. Reduce.

$1\frac{1}{3} \times 6 =$

$\frac{4}{3} \times \frac{6}{1} = \frac{8}{1} = 8$

1. Change each mixed number to an improper fraction.
2. Multiply the numerators. *Watch the sign!*
3. Multiply the denominators.
4. Reduce if possible.

TIP: *You can reduce first by dividing a numerator and denominator by a common factor. This is called cross cancellation.*

---

1. $\frac{3}{5} \times \frac{3}{4}$

2. $\frac{1}{2} \times \frac{5}{16}$

3. $6\frac{1}{2} \times 2\frac{2}{8}$

4. $\frac{4}{7} \times 7$

5. $2\frac{3}{4} \times 8\frac{4}{7}$

6. $-4\frac{4}{5} \times 3\frac{3}{4}$

7. $6 \times \frac{5}{12}$

8. $-7 \times 2\frac{5}{8}$

9. $\frac{6}{5} \times \frac{5}{6}$

10. $-7\frac{1}{7} \times \frac{-3}{4}$

11. $5 \times 2\frac{4}{5}$

12. $-4\frac{5}{6} \times 2\frac{2}{3}$

13. $1\frac{3}{4} \times \frac{-4}{5}$

14. $-3\frac{1}{10} \times 3\frac{1}{3}$

15. $-5\frac{3}{5} \times 2\frac{2}{8}$

16. $14 \times 2\frac{1}{7}$

17. $\frac{4}{7} \times 2\frac{1}{3} \times \frac{5}{12}$

18. $\frac{5}{8} \times 5 \times \frac{4}{5}$

19. $\frac{-6}{7} \times \frac{1}{4} \times \frac{2}{5}$

20. $8\frac{7}{9} \times 3 \times 5\frac{1}{2}$

 THINK ABOUT IT!

21. Evaluate each expression if $x = \frac{2}{3}$, $y = -1\frac{1}{2}$, and $z = 3\frac{3}{4}$.

     a. y^2 b. xyz   c. $z^2(x + y)$

Name _____

Rational Numbers: Dividing Fractions and Mixed Numbers

Find each quotient. Reduce.

$$-3\frac{7}{10} \div 2\frac{1}{2}$$

$$\frac{-37}{10} \div \frac{5}{2}$$

$$\frac{-37}{10} \times \frac{2}{5} = \frac{-74}{50} = -1\frac{12}{25}$$

1. Write the mixed numbers (or whole numbers) as improper fractions.
2. To divide fractions, flip the second one and then multiply.
3. Reduce.

TIP: *Remember that a whole number can be written as a fraction by placing it over one.* $6 = \frac{6}{1}$

1. $\dfrac{3}{5} \div \dfrac{1}{5}$

2. $\dfrac{-3}{10} \div \dfrac{8}{5}$

3. $\dfrac{1}{9} \div \dfrac{-4}{3}$

4. $4 \div 1\dfrac{1}{4}$

5. $7 \div \dfrac{7}{8}$

6. $-6\dfrac{1}{8} \div 4\dfrac{2}{3}$

7. $-3\dfrac{1}{5} \div 5$

8. $2\dfrac{1}{2} \div \dfrac{-3}{4}$

9. $-3\dfrac{1}{5} \div 4\dfrac{2}{5}$

10. $5 \div -1\dfrac{1}{3}$

11. $12\dfrac{1}{4} \div \dfrac{-14}{3}$

12. $-7\dfrac{1}{2} \div 1\dfrac{1}{5}$

13. $2 \div \dfrac{-1}{3}$

14. $\dfrac{5}{7} \div \dfrac{-1}{14}$

15. $5\dfrac{5}{6} \div 2\dfrac{1}{3}$

16. $12 \div 1\dfrac{1}{4}$

17. $-7 \div \dfrac{1}{2}$

18. $-4\dfrac{3}{5} \div 3\dfrac{1}{5}$

48

Rational Numbers: Solving Equations   Name _____

Solve and check each equation.

$$\frac{5}{6}x = \frac{7}{12}$$

$$\frac{5}{6}x \div \frac{5}{6} = \frac{7}{12} \div \frac{5}{6}$$

$$x = \frac{7}{10}$$

$$\frac{5}{6} \times \frac{7}{10} = \frac{7}{12}$$
$$\frac{35}{60} = \frac{7}{12}$$
$$\frac{7}{12} = \frac{7}{12} \checkmark$$

1. Look at what has been done to the variable.

2. Undo it by using the inverse (opposite) operation on both sides of the equation.

3. Check your answer by plugging it back into the equation to see if it makes the equation true.

---

1. $\dfrac{-3}{4}x = 2\dfrac{1}{2}$

2. $6y = \dfrac{6}{10}$

3. $\dfrac{5}{6}m = 4$

4. $\dfrac{5}{6}y = 6$

5. $\dfrac{y}{3} = \dfrac{4}{3}$

6. $\dfrac{-9}{16}x = \dfrac{-5}{8}$

7. $\dfrac{x}{-10} = \dfrac{4}{5}$

8. $7b = \dfrac{-3}{5}$

9. $-1\dfrac{2}{3}x = \dfrac{6}{5}$

10. $2\dfrac{1}{2}m = 3\dfrac{3}{4}$

11. $\dfrac{x}{12} = 2\dfrac{3}{10}$

12. $-3\dfrac{4}{5} = 2\dfrac{5}{6}m$

13. $3\dfrac{3}{10} = 4\dfrac{3}{5}x$

14. $-\dfrac{7}{8}x = \dfrac{7}{12}$

15. $-5n = \dfrac{3}{5}$

Ratio and Proportion: Ratios & Equal Ratios

Part I: Express each ratio as a fraction in simplest form.

$$8 \text{ to } 12$$
$$\frac{8}{12} = \frac{2}{3}$$

A ratio is a comparison of two numbers. It is often written as a fraction in simplest form.

1. 2 to 10

2. 34 : 12

3. 8 to 14

4. 18 to 30

5. 10 out of 26

6. 7 : 35

7. 24 hours : 60 hours

8. 36 to 64

9. 10 ft to 90 ft

Part II: Determine whether each pair of ratios is equal. Write **yes** or **no.**

$$15 \qquad 16$$
$$\frac{5}{8} \bowtie \frac{2}{3}$$
(no)

1. Cross multiply.
2. If the cross products are equal, the ratios are equal.

10. $\frac{6}{8}$, $\frac{22}{28}$

11. $\frac{4}{15}$, $\frac{3}{7}$

12. $\frac{8}{15}$, $\frac{20}{45}$

13. $\frac{3}{8}$, $\frac{6}{16}$

14. $\frac{3}{8}$, $\frac{12}{32}$

15. $\frac{7}{15}$, $\frac{20}{35}$

16. $\frac{5}{7}$, $\frac{2.1}{3.5}$

17. $\frac{8}{9}$, $\frac{6}{17}$

18. $\frac{12}{30}$, $\frac{3}{8}$

19. $\frac{6}{8}$, $\frac{3}{4}$

20. $\frac{4}{9}$, $\frac{28}{60}$

21. $\frac{4}{6}$, $\frac{14}{21}$

Ratio & Proportion: Proportions

Name _____

Solve each proportion.

$$\frac{3}{4} = \frac{n}{48}$$

A proportion is two equal ratios. To solve:
1. Find the cross products.

$$4 \times n = 3 \times 48$$
$$n = 144 \div 4$$
$$n = 36$$

2. Division reverses multiplication to solve for n.

---

1. $\frac{3}{6} = \frac{n}{24}$

2. $\frac{5}{2} = \frac{14}{n}$

3. $\frac{1}{2} = \frac{n}{34}$

4. $\frac{12}{27} = \frac{8}{x}$

5. $\frac{2}{7} = \frac{n}{49}$

6. $\frac{2.5}{4} = \frac{10}{x}$

7. $\frac{2}{5} = \frac{x}{35}$

8. $\frac{8}{12} = \frac{n}{3}$

9. $\frac{x}{5} = \frac{12}{15}$

10. $\frac{y}{24} = \frac{15}{60}$

11. $\frac{7}{8} = \frac{n}{56}$

12. $\frac{2.6}{13} = \frac{8}{h}$

13. $\frac{5}{9} = \frac{n}{5.4}$

14. $\frac{4}{9} = \frac{36}{x}$

15. $\frac{1}{2} = \frac{c}{7}$

16. $\frac{x}{42} = \frac{5}{140}$

17. $\frac{3}{48} = \frac{n}{72}$

18. $\frac{p}{20} = \frac{120}{150}$

19. $\frac{3}{w} = \frac{18}{24}$

20. $\frac{40}{8} = \frac{150}{c}$

21. $\frac{8}{15} = \frac{n}{105}$

22. $\frac{7}{8} = \frac{n}{72}$

23. $\frac{n}{36} = \frac{2}{4}$

24. $\frac{30}{40} = \frac{n}{4}$

 THINK ABOUT IT!

25. Solve each proportion for x.

a. $\frac{5}{6} = \frac{25}{x+1}$

b. $\frac{3}{7} = \frac{x+1}{42}$

c. $\frac{1}{5} = \frac{x+2}{25}$

51

Ratio & Proportion: Rates

Name _____

Express each rate as a unit rate. Round answers to the nearest tenth, or the nearest penny.

> **Rate:** a ratio of two measurements with different units
> **Unit Rate:** a rate in which the denominator is 1

$1.99/200 napkins $1.99 ÷ 500 = $0.00995 per 1 napkin = approx. $0.01 per napkin

---

1. 240 people/15 train cars

2. 280 tickets/2 hours

3. 325 miles/13 gallons

4. $.89/dozen eggs

5. 810 miles/6 days

6. 18 people in 4 cars

7. 9 pounds in 3 weeks

8. $5.97/3 pounds

9. $2996/14 day trip

10. 35 people/7 rows

11. $29.00 for 4 tapes

12. 315 miles/5 hours

13. $144/3 outfits

14. 1,425 students/57 teachers

15. 9,000 words/2.5 hours

16. $36.00 for 6 tapes

 THINK ABOUT IT!

17. How long would it take to lay 12 rows of 16 bricks at a rate of 4 bricks per minute?

MATH FACTS
A branch of math called "Knot Theory" is used to identify viruses in the human body. A virus manipulates human cells by changing a cell's DNA into a "knot-like" shape. Each particular virus has a unique "knot shape" much like people have unique fingerprints.

Ratio & Proportion: Percent

Name _____

Part I: Write each ratio as a percent.

$48:100 = 48\%$ *A percent is a special ratio that compares a quantity to 100.*

$\dfrac{83}{100} = 83\%$

1. 88:100
2. $\dfrac{53}{100}$
3. $\dfrac{25}{100}$
4. 36:100

5. 1 to 100
6. 99:100
7. 27 to 100
8. $\dfrac{63}{100}$

9. 15 to 100
10. 43:100
11. $\dfrac{21}{100}$
12. 71:100

Part II: Write each percent as a ratio (use fractions in lowest terms).

13. 60%
14. 17%
15. 5%
16. 20%

17. 100%
18. 35%
19. 8%
20. 5¾%

21. 9½%
22. 63%
23. 92%
24. 44%

 THINK ABOUT IT!

25. Use each diagram below to show 25%.

   a.

   b.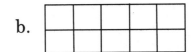

   c.

   d.

Ratio & Proportion: Percents & Decimals   Name _____

Part I: Express each decimal as a percent.

0.06
0.06 × 100 =
6%

To change a decimal to a percent, multiply by 100 and add the percent sign.

TIP: *When you multiply by 100, it moves the decimal point two places to the right. Simply moving the decimal point two places to the right changes the decimal to a percent.*

---

1. 0.508

2. 0.01

3. 0.4

4. 0.55

5. 2.4

6. 0.07

7. 0.023

8. 0.75

9. 0.3

10. 4.28

11. 0.245

12. 0.006

Part II: Express each percent as a decimal.

45%
0.45

To change a percent to a decimal, move the decimal point two places to the left.
2% = 0.02   80% = 0.8

---

13. 72%

14. 4%

15. 17%

16. 90%

17. 12.1%

18. 200%

19. 25%

20. $\frac{1}{4}$%

21. $\frac{3}{4}$%

22. $10\frac{1}{2}$%

23. 450%

24. 80%

Ratio & Proportion: Problem Solving

Solve each problem.

1. The ratio of boys to girls in Sally's school is 4:5. If there are 304 boys in the school, how many students are in the school?

1. _____

2. The ratio of fans cheering for the home team vs. the visiting team is 7 to 2. If there are 18,000 fans at the game, approximately how many are cheering for the home team?

2. _____

3. What is the hourly rate of a plumber who charged $380 for an 8 hour job?

3. _____

4. Anne's snack mix consists of peanuts and raisins in a ratio of 7:4. How many raisins does she have if there are 35 peanuts?

4. _____

5. A cookie recipe calls for 3 cups of flour for every 4 dozen cookies. How much flour would you need to bake 72 cookies?

5. _____

6. Keri's softball team won 24 games and lost 4 during the season. About what percent of the games did her team win?

6. _____

7. Hamburger patties are on sale at the butcher for $1.25/lb. How many pounds of hamburger can you buy for $5.00?

7. _____

Using Percent: Percent of a Number

Name _____

Use a proportion to solve each problem. Round answers to the nearest tenth.

What number is 40% of 20?

$$\frac{x}{20} = \frac{40}{100}$$

$100x = 800$

$x = 8$

Answer: **8** is 40% of 20.

Percent Proportion

$$\frac{\text{Part}}{\text{Whole}} = \frac{\%}{100}$$

1. Identify the part, whole, and/or percent.

2. Plug the numbers into the proportion and solve for the missing piece (part, whole or %).

---

1. 30 is what percent of 150?

2. Find 9% of 24.5.

3. 2.5% of 30 is what number?

4. What number is 4% of 60?

5. 9 is 15% of what number?

6. Find 60% of $90\frac{1}{2}$.

7. 77 is what percent of 154?

8. What number is 90% of 50?

9. Find 250% of 20.

10. $\frac{1}{2}$% of 200 is what number?

11. 65 is what percent of 50?

12. 4.5% of 60 is what number?

13. What number is 30% of 412?

14. 3.9 is what percent of 3.9?

Using Percent: Simple Interest

Find the interest to the nearest cent.

principal: $800
    rate: 18%
    time: 1 year

$$Interest = Principal \times Rate \times Time$$
$$I = prt$$

$$I = 800 \cdot 0.18 \cdot 1 = \$144$$

TIP: *Time should be expressed in years. For example: 6 months equals $\frac{6}{12} = 0.5$ year.*

---

1. principal: $4,200
    rate: 9%
    time: 3 years

2. principal: $340
    rate: 12%
    time: 1.5 years

3. principal: $200
    rate: 6%
    time: 2 years

4. principal: $123
    rate: 16%
    time: 3 months

5. principal: $325
    rate: 18.5%
    time: 1 years

6. principal: $1,200
    rate: 19%
    time: 9 months

7. principal: $2,250
    rate: 7%
    time: 3 years

8. principal: $175
    rate: 12%
    time: $1\frac{1}{4}$ years

9. principal: $88.50
    rate: $6\frac{1}{2}$%
    time: 15 months

10. principal: $1,000
    rate: $20\frac{1}{2}$%
    time: 1 year

Name _____

Using Percent: Percent of Increase or Decrease

Find the percent of increase or decrease and state whether it is an increase or decrease. Round to the nearest whole percent.

old: 50 (original)
new: 75
75 − 50 = 25 ←——————— 1. Find the amount of increase or decrease.
                      2. Write a proportion:
$\frac{25}{50} = \frac{x}{100}$ ← $\frac{\text{change}}{\text{original}} = \frac{x\%}{100}$
2,500 = 50x
x = 50% ←——————— 3. Solve the proportion to find the percent
                      of change.

_____

1. old: $10   2. old: 72   3. old: 40
    new: $12   new: 36   new: 34

4. old: 200   5. old: $245   6. old: $7.80
    new: 260   new: $383   new: $9.20

7. old: $400   8. old: 50   9. old: 68
    new: $356   new: 28   new: 79

10. old: $0.85   11. old: 240   12. old: $320
    new: $1.15   new: 268   new: $350

13. old: 70   14. old: 40   15. old: 7.5
    new: 45   new: 15   new: 10

 THINK ABOUT IT!

16. Last month, the manager of Portman's Music Store decreased all prices of CD's by 10%. This month he increased all prices of CD's by 5%. If you bought a CD last month for $13.95, what would you have paid for it this month?

Using Percent: Problem Solving

Solve each problem.

1. Lakeside High School has a graduating
   class of 325 students. If 60% of the students
   are going to college, how many Lakeside
   High students will be attending college?

   1. _____

2. Of the 350 fans surveyed at a football game,
   250 said it was their first time at a game.
   What percentage saw their first game that
   day?

   2. _____

3. A science test had 30 problems. If Rachel
   answered 22 correctly, what percent is that?

   3. _____

4. A history test has a total of 150 points.
   Armen received 110 points. If 70% is
   considered passing, did Armen pass the
   test?

   4. _____

5. Pete made 12 field goals out of 18
   attempted. What percent of field goals
   attempted did Pete make?

   5. _____

6. If Sarah invested $2,500 in a fund that
   earned 12% interest each year, how much
   would her fund be worth after 3 years?

   6. _____

7. Bill's computer cost $2,000 five years ago. He
   recently purchased a new computer for
   $1,350. What is the percent of change in the
   prices?

   7. _____

8. A video game cost $27.75 two years ago.
   Today, the same video game is $31.95.
   By what percentage did the price
   increase?

   8. _____

Geometry: Perimeter

Part I: Find the perimeter.

Perimeter: distance around a figure

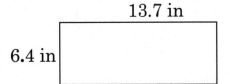

13.7 in

6.4 in

$P = 2(6.4 + 13.7) = 40.2$ in

1.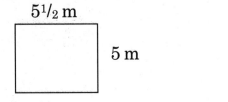

$5^1/_2$ m

5 m

2.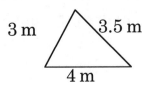

3 m 　 3.5 m

4 m

3.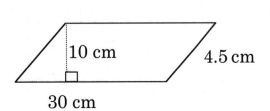

10 cm 　 4.5 cm

30 cm

4.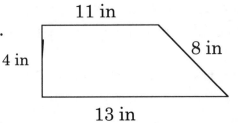

11 in

4 in 　 8 in

13 in

Part II: Given the perimeter of each figure, find the unknown length(s).

5.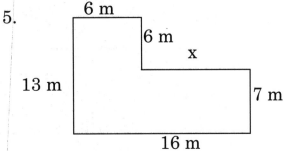

6 m

6 m

x

13 m

7 m

16 m

Perimeter = 58 m

6.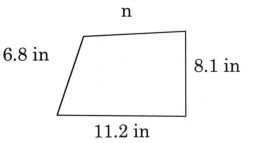

n

6.8 in 　 8.1 in

11.2 in

Perimeter = 35.6 in

7.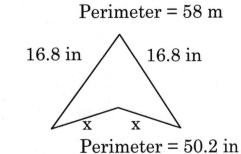

16.8 in 　 16.8 in

x 　 x

Perimeter = 50.2 in

8.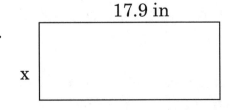

17.9 in

x

Perimeter = 50.8 in

Geometry: Angles

Name _____

Part I: Classify each angle as <u>acute</u>, <u>right</u>, <u>obtuse</u>, or <u>straight.</u>

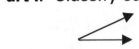

Acute Angle   Right Angle   Obtuse Angle   Straight Angle
< 90°   = 90°   90° < x < 180°   = 180°

1.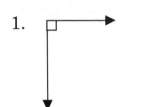

2.

3.

4.

5.

6.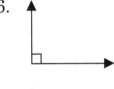

Part II: Write and solve an equation to find the following.

Complementary angles
Two angles are
complementary if the
sum of their measures = 90°.

Supplementary angles
Two angles are
supplementary if the
sum of their measures = 180°.

the complement of a 21° angle
$$x + 21 = 90$$
$$x + 21 - 21 = 90 - 21$$
$$x = 69°$$

7. the supplement of a 123° angle

8. the complement of 57° angle

9. the complement of a 43° angle

10. the supplement of a 45° angle

11. the complement of a 19° angle

12. the supplement of a 93° angle

13. the supplement of a 110° angle

14. the complement of a 38° angle

Geometry: Classifying Triangles

Name _____

Part I: Classify each triangle by its sides and angles.

By sides:

<u>Equilateral</u>

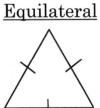

3 equal sides

<u>Isosceles</u>

2 equal sides

<u>Scalene</u>

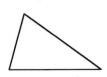

no equal sides

By angles:

<u>Acute</u>

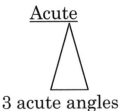

3 acute angles

<u>Obtuse</u>

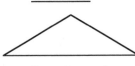

1 obtuse angle
2 acute angles

<u>Right</u>

1 right angle
2 acute angles

1.

2.

3.

4.

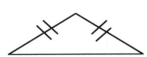

Part II: Find the missing angle in each triangle.

$x + 65 + 55 = 180$

$x = 60°$

The sum of the measures of the angles of any triangle is 180°.

5.

6. 31°

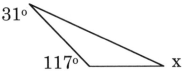

117°   x

7.

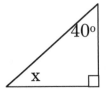

8.

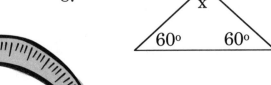

62

Geometry: Pythagorean Theorem

Name _____

Part I: Use the *Pythagorean Theorem* to find the length of each hypotenuse given the lengths of the legs. Round answers to the nearest tenth.

3 ft, 6 ft.

$3^2 + 6^2 = c^2$
$9 + 36 = c^2$
$45 = c^2$
$\sqrt{45} = c$
$\boxed{c = 6.7 ft}$

The *Pythagorean Theorem* states:

<u>In a right triangle</u>, the square of the measure of the hypotenuse is equal to the sum of the squares of the measure of the legs.

legs   hypotenuse

$a^2 + b^2 = c^2$

legs   hypotenuse

legs   hypotenuse   legs

1. 9 m, 12 m

2. 3 cm, 4 cm

3. 10 in, 16 in

4. 8 ft, 15 ft

5. 10 m, 23 m

6. 12 in, 5 in

7. 12 ft, 20 ft

8. 7 cm, 9 cm

9. 7 ft, 24 ft

10. 6 mm, 15 mm

11. 9 in, 40 in

12. 9 ft, 11 ft

Part II: Given the following lengths, determine whether each triangle is a right triangle. Write **yes** or **no.** The first two lengths in each problem are the "legs" while the third is the "hypotenuse."

13. 9 ft, 12 ft, 15 ft

14. 7 in, 10 in, 11 in

15. 3 cm, 4 cm, 5 cm

16. 9 in, 11 in, 17 in

17. 9 yd, 40 yd, 41 yd

18. 7 m, 24 m, 25 m

19. 8 ft, 15 ft, 17 ft

20. 2 in, 3 in, 4 in

21. 20 m, 25 m, 30 m

Geometry: Circumference & Area of Circles

Find the circumference and area of each circle. Use 3.14 for π. Round to the nearest hundredth.

Circumference: Distance around

$C = \pi \times d$
$= 3.14 \times 12$
$= 37.68 \text{ in}$

Area: What's inside

$A = \pi \times r^2$
$= 3.14 \times 5^2$
$= 78.5 \text{ m}^2$

TIP: *Remember that the diameter is twice the radius. d = 2r.*

---

1. 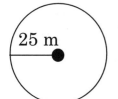 C=
A=

2. 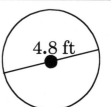 C=
A=

3. 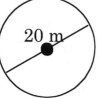 C=
A=

4. 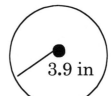 C=
A=

5.   C=
A=

6. 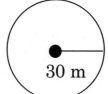 C=
A=

THINK ABOUT IT!

7. Given the area or circumference of the following circles, find the radius and diameter. Use π = 3.14
   a. c = 40.82 cm   b. A = 15.1976 in²   c. A = 452.16 cm²

Integers: Problem Solving

Solve each problem.

1. Raymond's score at the end of a game was 60. About halfway through the game, he had a score of −25. How many points did he score during the last half of the game?

1. _____

2. On first down, Barry's football team lost 3 yards. On second down, they gained 6 yards. On third down, they lost 2 yards. On fourth down, they gained 12 yards. What was their total gain or loss after 4 downs?

2. _____

3. The quotient of two integers is ⁻28. The divisor is ⁻7. What is the other dividend?

3. _____

4. The temperature was recorded at 52°F. If it rose 4°, fell 6°, rose 8°, then fell 12°, what was the temperature after these changes?

4. _____

5. An elevator went down 2 floors, up 10 floors, down 6 floors, and up 3 floors. It stopped on the 22nd floor. On what floor did it start?

5. _____

6. Sally's bank account is overdrawn by $30. How much does she need to deposit to have a balance of $100?

6. _____

7. The outside temperature was 15°F. With the windchill, it made the temperature feel as though it were −8°F. Find the difference between the real temperature and the apparent temperature.

7. _____

Integers: Multiplying & Dividing

Find each product or quotient.

When the signs are the **same** (both positive or negative) the answer will be **positive.**	When the signs are **different** (one positive and one negative) the answer will be **negative.**
$-8 \cdot -9 = 72$ $60 \div 5 = 12$	$-63 \div 9 = -7$ $5 \cdot -9 = -45$

1. $-5 \cdot -3$

2. $-48 \div -3$

3. $-3 \cdot -8$

4. $-56 \div 8$

5. $-9 \cdot 9$

6. $-18 \cdot -3$

7. $72 \div -12$

8. $7 \cdot -8$

9. $-12 \cdot 35$

10. $-64 \div 16$

11. $-98 \div -14$

12. $0 \cdot -37$

13. $91 \div -13$

14. $54 \cdot -23$

15. $-24 \cdot -52$

16. $240 \div -6$

17. $86 \cdot 45$

18. $-288 \div -24$

19. $(6 \cdot -5) \cdot -4$

20. $(-52 \div 4) \cdot 3$

21. $-7 \cdot (6 \cdot -4)$

22. $24 \div (-6 \cdot -4)$

THINK ABOUT IT!

23. Find two integers for x and y to make both statements true.

$x + y = -5$ $\qquad$ $xy = 6$

Computation & Estimation: Order of Operations

Evaluate each expression. Remember to use the correct order of operations.

Order of operations:
1. Work inside all grouping symbols first.
2. Evaluate exponents.
3. Next, multiply and divide, in order *from left to right.*
4. Last, add and subtract, in order *from left to right.*

$$(25 - \underline{4^2}) \times 2$$
$$\underline{(25 - 16)} \times 2$$
$$9 \qquad \times 2 = 18$$

---

1. $12 \div 3 + 12 \div 4$

2. $21 \div 7 + 4 \cdot 11$

3. $(21 \div 7 + 4) \cdot 11$

4. $15 \div 5 \cdot 3$

5. $15 \div (5 \cdot 3)$

6. $96 \div 12 \, (4) \div 2^2$

7. $40 \cdot 2 - 6 \cdot 11$

8. $(24 \div 2^3) \cdot 3$

9. $3[6 \, (12 - 3)] - 17$

10. $7[(12 + 5) - 3(4)]$

11. $6 + 5^2 - 2$

12. $[15 - (4 \cdot 2)] + 9$

13. $144 \div 16 \cdot 9 \div 3$

14. $\dfrac{72 + 12}{35 + 7}$

15. $\dfrac{86 - 11}{9 + 6}$

 THINK ABOUT IT!

16. Write an expression which has multiplication and addition where you should add first.

Computation & Estimation: More Exponents

Simplify. Give answers in exponent form.

Multiplying Powers with Like Bases	*Dividing Powers with Like Bases*
Add the exponents. Use the sum as the exponent together with the base.	Subtract the exponents. Use the difference as the exponent together with the base.
$2^3 \cdot 2^4 = 2^{3+4} = 2^7$	$\dfrac{5^5}{5^3} = 5^{5-3} = 5^2$

1. $3^2 \cdot 3^4$

2. $4^4 \cdot 4^6$

3. $9^7 \cdot 9^3$

4. $\dfrac{2^6}{2^3}$

5. $\dfrac{10^4}{10}$

6. $\dfrac{3^8}{3^2}$

7. $5^5 \cdot 5^8$

8. $6^2 \cdot 6^4 \cdot 6^5$

9. $7 \cdot 7^5$

10. $\dfrac{4^3}{4^2}$

11. $\dfrac{8^3}{8}$

12. $\dfrac{3^5}{3^4}$

13. $a^2 \cdot a^4$

14. $x^2 \cdot x \cdot x^3$

15. $y^3 \cdot y^5$

16. $\dfrac{n^6}{n^4}$

17. $\dfrac{b^2}{b}$

18. $\dfrac{c^3}{c^2}$

 THINK ABOUT IT!

19. Evaluate each of the following expressions if $a = 3$, $b = 2$, and $c = 4$.

    a. $a^3 \cdot a^2$ b. $5^b + 5^a$ c. $10^b + a$ d. $c^5 \div c^2$

Integers: Subtracting Integers

Find each difference.

To subtract integers, add the opposite.

$20 \underline{- (^-7)}$
$20 \underline{+ 7} = 27$

$^-12 \underline{\ - \ 6}$
$^-12 \underline{+ (^-6)} = ^-18$

$^-5 \underline{- (^-8)}$
$^-5 \underline{+ 8} = 3$

TIP: *Never change the sign of the first integer, just the second.*

---

1. $3 - 10$

2. $^-6 - 9$

3. $^-12 - 3$

4. $6 - 13$

5. $24 - (^-16)$

6. $23 - 14$

7. $^-5 - 3$

8. $7 - 35$

9. $51 - (^-11)$

10. $^-42 - 38$

11. $^-18 - (^-18)$

12. $17 - (^-6)$

13. $35 - (^-17)$

14. $15 - (^-28)$

15. $^-63 - 14$

16. $17 - (^-2)$

17. $^-24 - 19$

18. $^-41 - 12$

19. $^-9 - (^-9)$

20. $^-67 - 1$

21. $13 - (^-18)$

 THINK ABOUT IT!

22. Replace x, y, and z with integers to show that $(x - y) - z = x - (y - z)$ is not true for all integers.

MATH FACTS
Math is much more than exercises you are practicing in this book. The modern science of mathematics is used by people in all walks of life — from astronomers and ocean ship captains to computer game designers and sports trainers — math is the language of life!

Integers: Adding Integers

Name _____

With the Same Sign	**With Different Signs**
The sum of two positive integers is positive; the sum of two negative integers is negative.	Subtract their absolute values and use the sign of the original integer with the greater absolute value.
$^-6 + {}^-4 = {}^-10$	$^-8 + 16 = 8$
$23 + 32 = 55$	$\mid 16 \mid - \mid {}^-8 \mid =$
	$16 - 8 = 8$

Find each sum.

1. $^-10 + 3$

2. $5 + 31$

3. $(^-5) + (^-13)$

4. $(^-20) + (^-13)$

5. $^-72 + 88$

6. $^-27 + 45$

7. $33 + (^-48)$

8. $^-14 + 13$

9. $47 + (^-63)$

10. $^-28 + (^-13)$

11. $^-35 + (^-9)$

12. $16 + 18$

13. $^-12 + 14 + 8$

14. $7 + (^-11) + 32$

15. $^-18 + (^-23) + 10$

16. $^-31 + 19 + 4$

17. $^-17 + (^-10) + 18$

18. $34 + (^-34)$

19. $26 + (^-13 + {}^-13)$

20. $^-53 + 20 + (^-7)$

21. $^-18 + 18 + 35$

 THINK ABOUT IT!

22. Add each sum mentally.
   a. $-47 + 36 + 278 + (-35) + 199 + 48 + (-278)$
   b. $23 + (-17) + 31 + (-22) + (-32) + 17$

Integers: Integers & Absolute Value   Name _____

Part I: Write an integer for each exercise.

An ***integer*** is the set of whole numbers and their opposites.

If ⁻15 represents a loss of 15 pounds, write an integer for a gain of 15 pounds.
Answer: 15

TIP: *A plus sign is not needed to write a positive integer; however, a negative sign is required for a negative.*

---

1. An altitude of 1,000 feet

2. 8 inches taller

3. A loss of $150

4. An increase of 10 degrees

5. A profit of $100

6. A withdrawal of $20

7. The opposite of 23

8. The opposite of –35

Part II: Find each absolute value.

The ***absolute value*** of a number is its distance from zero. The following symbol is used when asked to find the absolute value: | |. (Two straight lines surrounding the number.)

| –25 | = 25   –25 is 25 places from zero so its ***absolute value*** is 25.
| 18 – 7 | = 11   Subtract the numbers first, then find the ***absolute value.***

9. | 5 |

10. | –6 |

11. | 12 |

12. | –4 |

13. | 7 |

14. | 0 |

15. | –11 |

16. | 15 |

17. | ⁻33 |

18. | 12 + 7 |

19. | 60 – 31 |

20. | 21 – 14 |

Computation & Estimation: Problem Solving

Solve each problem.

1. Cheryl bought items at the grocery store for the following prices: $1.39, $.58, $1.19, and $2.89. Estimate the total cost Cheryl spent to the nearest dollar.

1. _____

2. Sam bought a bicycle that cost $452.16. If he pays for the bike in 12 installments, estimate the cost of each payment to the nearest dollar.

2. _____

3. There were 3,542 runners entered in the annual Labor Day road race. If 1,054 runners did not finish the race, estimate the number of runners who did finish to the nearest hundred.

3. _____

4. Ray's car gets an average of 28.5 miles per gallon. The gas tank holds 13.2 gallons of gasoline. Approximately how far can Ray drive on a full tank of gasoline?

4. _____

5. Insert parentheses in the following equations to make each sentence true:
   a. $18 \div 3 + 6 = 2$
   b. $15 \div 21 - 18 - 4 = 1$
   c. $54 + 24 \div 3 - 10 = 16$
   d. $24 \div 2 - 4 + 8 = 0$

5. _____

6. To find the volume of a cube, you find the product of its length, width, and height. Using exponents, write an expression representing the volume of a cube if it measures s units on each side.

6. _____

Geometry: Area Of Rectangles & Parallelograms   Name _____

Part I: Find the area of each figure. *Assume all units are meters.*

Rectangle
Area = Length × width

A = l x w

5 | 15

$5 \cdot 15 = 75$ m²

Parallelogram
Area = base × heigth

A = b x h

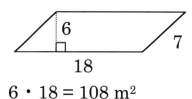

6 | 7 | 18

$6 \cdot 18 = 108$ m²

TIP: *In parallelograms the base and height are perpendicular.*

1.

12
7

2.

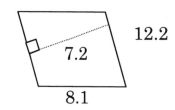

12.2
7.2
8.1

3.

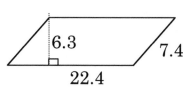

6.3
7.4
22.4

4.

18
38

5.

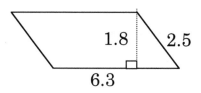

1.8   2.5
6.3

6.

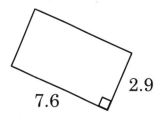

2.9
7.6

7. Parallelogram
   base = 13.3
   height = 12

8. Rectangle
   length = 23
   width = 5

9. Rectangle
   length = 22.8
   width = 19.4

10. Parallelogram
   base = 13.7
   height = 5.1

 THINK ABOUT IT!

11. A rectangle has a length of 15 cm and a perimeter of 44 cm. What is its area?

Geometry: Area of Triangles and Trapezoids

Part I: Find the area of each triangle. Round answers to the nearest tenth.

Area: measure of what is inside

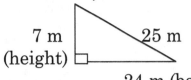

7 m
(height)   25 m

24 m (base)

b = base, h = height
A = ½ (b x h)
  = ½ (7 x 24)
  = 84 m²

TIP: *The base and height of a triangle will always be perpendicular.*

1.

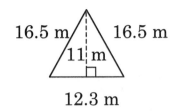

16.5 m   16.5 m
11 m

12.3 m

2.   3 ft.

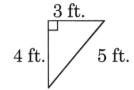

4 ft.   5 ft.

3. $5\frac{1}{5}$ cm

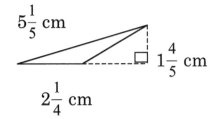

$1\frac{4}{5}$ cm

$2\frac{1}{4}$ cm

4.
4.7 cm

9.8 cm

3.2 cm

Part II: Find the area of each trapezoid.

5 in

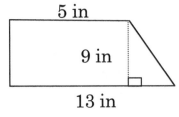

9 in

13 in

A = ½h (a + b), where **h** is the height
and **a** and **b** are the bases.

½(9)(5 + 13) = 81 in²

5.   9 in

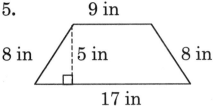

8 in   5 in   8 in

17 in

6.   7.6 m

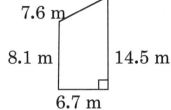

8.1 m   14.5 m

6.7 m

7. bases: $5\frac{1}{2}$ m, $6\frac{1}{3}$ m

height: 7 m

8. bases: 6.7 ft, 1.2 ft
    height: 0.34 ft

Geometry: Volumes of Prisms and Cylinders   Name _____

Find the volume of each prism or cylinder. Use 3.14 for π.

<u>Prisms</u>
V = (area of base)(height)

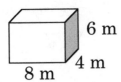

V = (8m)(4m)(6m) = 192 m³

<u>Cylinders</u>
V = (π r²)(height)

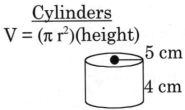

V = (π)(5²cm)(4 cm) = 314 cm³

TIP: *The base of a cylinder is a circle; the base of a prism is one of the two faces that are congruent. (In a triangular prism, the base would be a triangle.)*

1.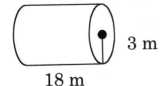

2. 7 in 6 in 11 in

3.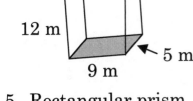

4. 14 cm 16 cm
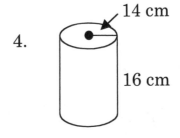

5. Rectangular prism
      1 = 25 cm
      w = .05 cm
      h = 15 cm

6. Cylinder
      r = 9 in
      h = 22 in

7. Cylinder
      r = 5 m
      h = 9 m

8. Rectangular prism
      1 = 7 cm
      w = 2.3 cm
      h = 1.2 cm

 THINK ABOUT IT!

9. A rectangular fish tank is 24 inches long, 30 inches high, and 18 inches wide. If the water in the tank is 26 inches high, what is the volume of the water in the tank?

Geometry: Volumes of Pyramids and Cones   Name _____

Find the volume of each pyramid or cone. Round answers to the nearest tenth.*

<u>Pyramid</u>

V = (¹/₃)(area of base)(height)

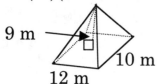

9 m

10 m

12 m

V = ¹/₃ (12 · 10)(9) = 360 m³

<u>Cones</u>

V = (¹/₃)(π r²)(height)

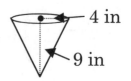

4 in

9 in

V = ¹/₃ π (4²)(9) = 150.7 in³

1.

s

s

s = 8 cm

h = 11 cm

2.

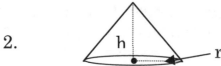

h

r

r = 12 m

h = 11 m

3.

5 m

9 m

h = 11 m

4.   r = 9 m

h = 16 m

5. rectangular pyramid

   1 = 10 m

   w = 8 m

   h = 19 m

6. cone

   r = 4 m

   h = 15 m

7. square pyramid

   s = 5 cm

   h = 19 cm

8. cone

   r = 5 m

   h = 10 m

 THINK ABOUT IT!

9. The volume of a sphere is V = ⁴/₃ π r³ . Use this formula to find the volume of the
   following spheres with each radius given. Express answers in terms of π.

   a. r = 15 cm

   b. r = 12 cm

   c. r = 9 cm

   d. r = 6 cm

*Use 3.14 for π.

Geometry: Surface Area

Name _____

Find the surface area of each prism or cylinder. Use 3.14 for π.

Surface area is the combined area of all the surfaces of a 3-dimensional figure.

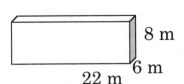

8 · 6 = 48 · 2 = 96
22 · 6 = 132 · 2 = 264
22 · 8 = 176 · 2 = 352
96 + 264 + 352 = 712 m²

1.
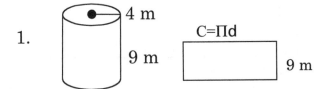

2.
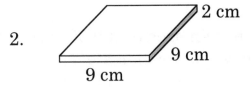

3.
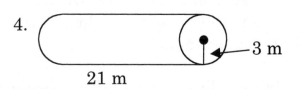

4.

5. Rectangular prism
    1 = 7 in
    w = 8 in
    h = 9 in

6. Cylinder
    r = 12 cm
    h = 30 cm

7. Cylinder
    r = 8 in
    h = 4 in

8. Cube
    s = 5 m

THINK ABOUT IT!

9. In your own words, explain how to find the surface area of a cylinder.

Geometry: Problem Solving

Solve each problem.

1. Sue wants to re-carpet her living room. Its dimensions are 14 ft x 20 ft. If Sue chooses carpet that costs $23/yd², how much will she pay to re-carpet the room? Remember that there are 3 ft. in a yard.

2. A trapezoid has bases of 9 m and 15 m. If its area is 72 m², what is its height?

3. A circle has an area of 200.96 cm². What is the length of its diameter?

4. The base of a 22-foot ladder is 8 feet from the house. Approximately how far above ground is the top of the ladder? (Hint: draw a diagram.)

5. Find the radius of a cylinder's base whose volume is 1356.48 ft³ and height is 12 ft.

6. A builder needs to purchase concrete for the foundation of a house. The concrete sells for $42 per cubic yard. How much will he pay for a foundation that is 46 ft by 30 ft by 6 in? Remember that there are 12 in. in a ft.

7. A triangle has an area of 63 m². If its height is 7 m, what is its base?

8. Write and solve an equation to find the measure of the third angle of a triangle if two angles measure 68° and 34°. What type of triangle is this?

9. If the radius of a circle doubles, by how much does the circumference increase?

1.

2.

3.

4.

5.

6.

7.

8.

9.

Probability & Statistics: Fundamental Counting Principle

In each situation, find the total number of outcomes.

> Amy has to wear a uniform to school. She can choose from 5 different kinds of shirts and 4 different kinds of pants/shorts/skirts. How many different combinations of tops and bottoms are there?
>
> Multiply the number of choices in each set to derive the number of possible combinations.
>
> 5 shirts × 4 pants/shorts/skirts = 20 combinations

1. Selecting 1 sweater from 9 different sweaters and 1 scarf from 6 different scarves.

2. Choosing an interior color and exterior color for a new car if there are 4 interior colors and 7 exterior colors to choose from.

3. Choosing the first two digits for a telephone number if you can use the digits 1-9 for the first number and 0-9 for the second.

4. Choosing from 3 different kinds of soups and 4 different kinds of salads for lunch.

5. Flipping a penny, nickel, dime and quarter.

6. Choosing from one of two history courses, three mathematics courses, two science courses, and five electives.

7. Choosing from four different sweaters, three different pairs of mittens, and three different hats.

Probability & Statistics: Permutations

Name _____

Find the number of permutations.

> *A **permutation** is an arrangement of items in a specific order.*
>
> Find the number of permutations for the digits 1, 2, 3, and 4. (In other words, how many different ways can 1, 2, 3, and 4 be arranged?)
>
> $$4! = 4 \times 3 \times 2 \times 1 = 24$$
>
> read "4 factorial"

---

1. Stacy, Kim, Iris

2. h, i, s, t, o, r, y

3. Jack, Queen, King, Ace

4. 5, 10, 15, 20, 25

5. Summer, Fall, Winter, Spring

6. See, Spot, run

7. numbers 0-9

8. m, a, t, h, e, m, a, t, i, c, s

9. tic, tac, toe

10. A, B, C, D, E

11. 0, 2, 4, 6, 8, 10

12. Larry, Moe, Curly

 THINK ABOUT IT!

13. How many different six-digit zip codes are possible if no digit is repeated?

MATH FACTS
Surveyors use the geometry of right triangles, which you are learning about in this book, to measure the height or altitude of mountain peaks and the depths of valleys and ravines.

Probability & Statistics: Combinations

Name _____

Find the number of combinations.

A ***combination*** is an arrangement of items in which order does not matter.

3 pictures from 5.

$$\frac{5 \cdot 4 \cdot 3}{3 \cdot 2 \cdot 1}$$ ◄——— Number of permutations of 5 pictures taken 3 at a time.
◄——— Number of permutations of 3 pictures.

$$\frac{60}{6} = \boxed{10}$$ There are 10 combinations of pictures.

---

1. 2 colors from
    red, orange, blue, green

2. 4 numbers from 10, 20, 30,
    40, 50

3. 2 ties from 8 ties

4. 4 books from a list of 7

5. 3 stamps from a book of 20

6. 3 numbers from numbers 1-10

7. 2 letters from X, Y, Z

8. 5 students from a class of 23

9. 2 shirts from 10 shirts

10. 2 students out of 5 candidates

11. 2 applications from a choice
     of 12

12. 9 baseball players from a team
     of 15

13. 5 boys from a group of 9

14. 2 movies from 6

 THINK ABOUT IT!

15. There are 12 dogs in the annual frisbee contest. How many ways can there be a
    winner and runner-up?

Probability & Statistics: Independent & Dependent Events

Independent Events	*Dependent Events*
To find the probability of independent events occurring, find the product of each of the events occurring.	Find the probability of the first event. With that in mind, find the probability of the second event, and so on. Finally, multiply the probabilities.
Tossing heads on a coin, then tails.	If there are 11 chips in a bag, 3 red and 8 white, find the probability of picking a white, and without replacing it, a red chip.
$\frac{1}{2} \times \frac{1}{2} = \left(\frac{1}{4}\right)$	$\frac{8}{11} \times \frac{3}{10} = \frac{24}{110} = \left(\frac{12}{55}\right)$

Select two cards at random from a standard deck of 52 playing cards, replacing the first card. Find the probability of each.

1. P (heart, then a spade)

2. P (10, then a jack)

3. P (King, then a King)

4. P (red card, then a black one)

5. P (red card, then a red card)

6. P (6 of hearts, then a 6 of spades)

7. P (club, then another club)

8. P (red 4, then a black 5)

Suppose you have a bag of 14 marbles: 2 are blue, 4 are red, 5 are yellow, and 3 are green. Once a marble is chosen, it is not replaced. Find the probability of each.

9. P (blue, then red)

10. P (red, then blue)

11. P (yellow, then blue)

12. P (green, then yellow, then red)

13. P (blue, then blue)

14. P (red, then green, then red)

Probability & Statistics: Problem Solving   Name _____

Solve each problem.

1. George is buying clothes to go back to
   school. If he buys three pairs of pants,
   four shirts, and three sweaters, how
   many days could he wear a different
   three-piece outfit without repeating any
   outfits?

2. A math quiz consists of 10 true-false
   questions. How many combinations of
   answers are possible?

3. There are eight runners in the 400-meter
   dash. In how many ways can the runners
   finish the race, assuming there are no
   ties?

4. There are 5 blue marbles, 6 red marbles,
   2 green marbles, 4 yellow marbles, and 1
   black marble. If you select one marble at
   random, what is the probability of the
   following:
   a. P (blue)
   b. P (red or yellow)
   c. P (not black)
   d. P (green)

5. Grayson has five single gray socks and
   eight single black socks in his drawer.
   What is the probability that he will pick
   a pair of gray socks if he choose one out
   of the drawer and then another?

6. Tanner needs to pack six pairs of shorts
   for camp. How many ways can he choose
   the shorts from the ten pairs that he has?

1.

2.

3.

4.

5.

6.

Probability & Statistics: Stem & Leaf Plots   Name _____

Part I: Make a stem & leaf plot for each set of data.

26, 31, 20, 19, 15
35, 29, 38, 12, 11

stem	leaf
1	1259 (11,12,15,19)
2	069 (20,26,29)
3	158 (31,35,38)

1. Use the tens place value for the stem.

2. Use the ones place value for the leaves. Put all leaves in order from least to greatest, using the same "stem" for all numbers sharing the same value in the tens place. For example, 26 would be listed as:

stem	leaf
2	6

---

1.   12, 11, 22, 32, 35, 45, 46, 14, 14, 16

2.   51, 62, 63, 71, 58, 79, 65, 83, 89, 72, 79

3.   8, 20, 31, 17, 18, 22, 19, 27, 22, 26, 9, 10

4.   91, 95, 87, 73, 92, 95, 85, 86, 87, 90

Part II: Use the following stem and leaf plot to answer the questions.

Test scores in Ms. Halpern's math class.

6	3
7	4558
8	1236699
9	334778
10	00

5. What is the lowest test score? highest? _____

6. How many students took the test? _____

7. Where did most of the test scores fall? _____

8. What is the median (middle) test score? _____

9. What is the mean (average)? _____

10. Were most of the test scores above average if average = 75%? _____

Probability & Statistics: Histograms

Name _____

Use the following histograms to answer each question.

A **histogram** is a special type of bar graph whose bars touch. This graph shows the distribution of data.

Test Scores for Mrs. Truett's English Class

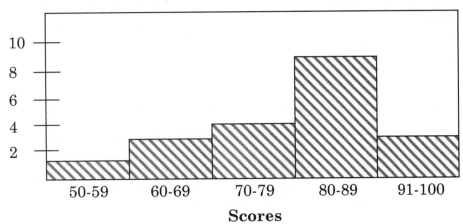

Scores

1. How many students took the test? _____
2. In what interval do most of the test scores fall? _____
3. From the information in the graph does it appear that Mrs. Truett's students studied for the test? _____
4. What information is not given on this graph? _____

_____

High Temperatures in December

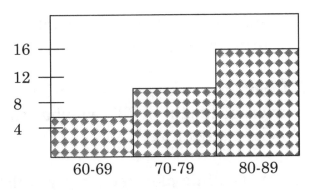

Temperatures

5. In which interval do most of the temperatures fall? _____

6. From this graph, can you tell what the highest temperature was in December? _____

7. In what type of climate would you say these temperatures were recorded? _____

Probability & Statistics: Line Plots

Name _____

Part I: Make a line plot (a way in which to display data) for each set of data.

4, 6, 5, 6, 10, 11
3, 5, 4, 11, 6, 4

1. First, draw a line. Determine your intervals.

```
    x   x
    x x x          x
  x x x x       x x
←—+—+—+—+—+—+—→
  2   4   6   8  10  12
```

2. Using an **x**, plot each number vertically.

TIP: *It may be helpful to check off each number as you plot it.*

_____

1. 40, 35, 45, 50, 40, 35, 55, 45, 50, 50

2. 72, 68, 67, 71, 72, 70, 72, 58, 65, 70

3. 20, 30, 20, 25, 35, 45, 40, 30, 45, 35

4. 500, 450, 550, 500, 400, 450, 500

Part II: Use the line plot below to answer each question.

Height (in inches) of players on
Shooters basketball team.

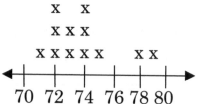

5. How tall, in feet, is the tallest player on the team? the shortest?

_____

6. Which height(s) most frequently represents a player? _____

7. If Shooter's plays a team with an average height of 73", who has the height advantage? _____

8. How is a line plot like a bar graph? How is it different? _____

_____

Graphing: Graphing on a Coordinate Plane   Name _____

Part I: Naming Points on a Graph. Name the coordinates of the points.
Use the coordinate system below. Remember: the first coordinate is the distance
from 0 on the **x-axis** and the second coordinate is the distance from 0 on the **y-axis**.

1. A

2. B

3. C

4. D

5. E

6. F

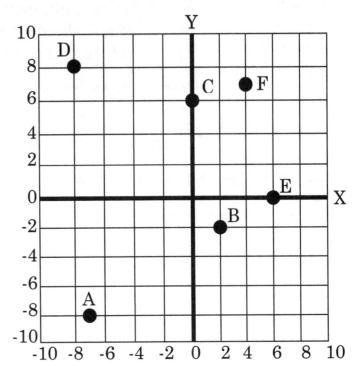

Part II: Graphing Points. On graph paper, draw a coordinate plane. Then
graph each set of points. Label each point.

7. G (⁻5, 4)

8. H (3, 4)

9. I (3, ⁻3)

10. J (0, 0)

11. K (⁻4, ⁻1)

12. L (⁻6, 0)

13. M (0, 8)

14. N (⁻1, 7)

 THINK ABOUT IT!

15. Graph the following pairs of points: (3, 2) and (2, 3); (⁻1, 3) and (3, ⁻1); (2, 5) and
(5, 2); (⁻4, 3) and (3, ⁻4); (⁻1, ⁻2) and (⁻2, ⁻1). What do you notice about the
graph? Make a generalization about the way the graphs of (x, y) and (y, x) are
related.

Graphing: Graphing on a Number Line

Solve each inequality. Then graph the solution on a number line.

$$2x - 3 < 1$$
$$2x - 3 + 3 < 1 + 3$$
$$2x < 4$$
$$2x \div 2 < 4 \div 2$$
$$x < 2$$

To graph the solution:
1. Draw a number line.
2. Shade the number line according to the solution of the inequality.

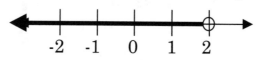

This graph shows that x < 2.

TIP: *For a less than or greater than sign, use a hollow circle. If an equals is added, fill in the circle.*

---

1. $x \geq {}^-2$

2. $x < {}^-1$

3. $^-x < {}^-4$

4. $^-2x \geq 4$

5. $x + 2 \leq {}^-3$

6. $x - 4 > 6$

7. $2x + 3 \leq {}^-5$

8. $x - 3 > {}^-2$

 THINK ABOUT IT!

9. Graph the solutions for the inequalities.
   a. $0 \leq x < 3$

   b. $^-1 < x < 3$

Graphing: Graphing Linear Equations

Name _____

Make a table of possible solutions for x = –2, –1, 0, 1 and 2, and graph each linear equation.

$y = x + 1$

Table of Solutions	
x	y
–2	–1
–1	0
0	1
1	2
2	3

Graph

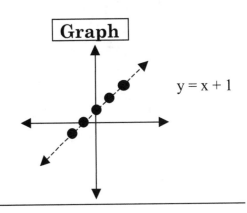

$y = x + 1$

---

1. $y = 2x - 1$

2. $y = 3x + 2$

3. $y = 3x - 4$

4. $y = -2x + 3$

5. $y = 5x + 7$

6. $y = -x + 6$

 THINK ABOUT IT!

7. Find an ordered pair that is a solution for both $x + y = 13$ and $x - y = -1$.

MATH FACTS
The Greeks used mathematics over 2,000 years ago to prove that the Earth was round. Until that time humankind assumed the Earth was flat and that you could sail to its edge and fall off. In this way, math has changed society's perceptions of its world.

More Equations & Inequalities: Combining Like Terms

Solve and check each equation.

$4x + 7x = 77$
$11x = 77$
$x = 7$

$4 \cdot 7 + 7 \cdot 7 = 77$
$28 + 49 = 77$
$77 = 77$ ✔

1. Combine like terms that have the same variable by adding or subtracting.
2. Solve for the variable.

3. Check your solution.

---

1. $-6y + 9y = 957$

2. $9x - 2x = -161$

3. $3x - 2x = 86$

4. $4y + 3y = 84$

5. $45 = 6n + 3n$

6. $125 = 30y - 25y$

7. $8x + 7x = 30$

8. $60 = 18y - 15y$

9. $4b + 12b = -48$

10. $-9b - 4b = 39$

11. $-3x + 9x = -54$

12. $6a + 9a = -75$

13. $-18x + 9x = 108$

14. $11y - 2y = 207$

15. $-65 = 8x - 3x$

16. $-14y + 8y = 126$

17. $288 = -24x + (-24x)$

18. $48x - 36x = -144$

19. $3.6y + 4.4y = 128$

20. $-x + 11x = 90$

21. $4.8n - 1.6n = 20.8$

More Equations & Inequalities: Simplifying to Solve

Solve and check each equation.

$$3(x - 7) + 2x = 44$$
$$3x - 21 + 2x = 44$$
$$5x - 21 = 44$$
$$5x = 65$$
$$\mathbf{x = 13}$$
$$3(13 - 7) + 2(13) = 44$$
$$18 + 26 = 44$$
$$44 = 44 ✔$$

1. Use the commutative, associative and/or distributive properties to simplify.
2. Solve for the variable.

3. Check your solution.

1. $6x + 2(x - 4) = 8$

2. $4(2n + 9) + 7n = ^-24$

3. $4(y - 3) + 12 = 64$

4. $5(x + 3) + 10 = ^-30$

5. $4n + 8 + 3(2n) = 128$

6. $6x + 4(x + 8) = 48$

7. $^-8y + 6(y + 7) = 2$

8. $10n + 5(n - 12) = 0$

9. $^-y + 3 + 7y = 93$

10. $12x - 2 + 5(3x) = 52$

11. $5(n + 4) + 6n = 97$

12. $^-2(3b) + 4(5b) + 12 = 82$

More Equations & Inequalities: Variables on Both Sides

Name _____

Solve and check each equation.

$$5y = 14 - y2$$
$$5y + 2y = 14 - 2y + 2y$$
$$7y = 14$$
$$y = 2$$
$$5 \cdot 2 = 14 - 2 \cdot 2$$
$$10 = 10 \ ✔$$

1. Move the variables to one side, and the numbers to the other.
2. Solve for the variable.

3. Check your solution.

---

1. $6x = 18 + 4x$

2. $4n = 7n + 12$

3. $7x + 24 = 5x$

4. $9y = 4y - 35$

5. $3b + 8 = 7b$

6. $12n + 18 = 6n$

7. $9n = 26 - 4n$

8. $15 - 8x = 7x$

9. $16n = 72 + 7n$

10. $^-5x + 144 = 7x$

11. $7x = 3x - 48$

12. $160 - 8y = 8y$

13. $4x = 30 - x$

14. $11y + 45 = 2y$

15. $6x = 9 + 9x$

16. $9x - 2 = 4x + 18$

17. $14n - 27 = 11n + 6$

18. $5a + 4 = 9a - 16$

19. $10 + x = 4x + 2x$

20. $9y + 8 = 11y + 4$

21. $3(a + 2) = 4 + 2a$

More Equations & Inequalities: Solving Multi-Step Inequalities

Solve each inequality.

$49 - 2x < 13 - 6x$
$49 < 13 - 4x$
$36 < {}^-4x$
$^-9 > x$
↓
$x < {}^-9$
Any number less
than $^-9$ is a *solution*.

To solve ***multi-step*** inequalities, use the same methods as when solving equations; **however**, when multiplying or dividing by a negative number, remember to change the sign of the inequality.

1. $^-5a - 9 > 21$

2. $8x + 7 \leq 71$

3. $3y - 5 < 13$

4. $^-7x + 4 > {}^-10$

5. $3 - 9x \leq 30$

6. $^-5c + 7 \geq -8$

7. $\dfrac{h}{5} + 36 \geq 51$

8. $\dfrac{x}{-3} - 4 > 27$

9. $\dfrac{y}{-2} - 12 \leq 11$

10. $2(n + 4) \leq 20$

11. $^-3(m - 2) < 15$

12. $^-21 \geq 3(k + 5)$

13. $36 + 7x > 8$

14. $6 + 9y < {}^-21$

15. $6 + 2y < 7y {}^-4$

16. $^-32 + 6a \leq {}^-2a$

17. $2y - 9 \geq {}^-5y + 12$

18. $^-2x + 3 < 5x - 4$

19. $2n + 5 > 6n + 4$

20. $^-6x + 3 < {}^-2x + 23$

21. $11y + 8 > 5y - 22$

Polynomials: Introduction

Part I: State whether each polynomial is a monomial, binomial, or trinomial.

y, 15, $2x^2$

$2x + 3$

$x^2 + 3x - 4$

A polynomial is a monomial or the sum or difference of monomials.

Monomial: an expression that consists of a number, variable, or product of numbers and variables.

Binomial: a polynomial with exactly two monomials.

Trinomial: a polynomial with exactly three monomials.

---

1. $7y^3 + 4x + 3$

2. $5b$

3. $-6r^2 + 2r$

4. $x^3 - y^2 - z$

5. 15

6. $4 + 3b - 6b^2$

7. $2x + 10$

8. $7y^2 + 2y - 3$

9. $6x^2$

Part II: Evaluate each polynomial for $w = -1$, $x = 2$, $y = -3$, and $z = 4$.

10. $4w^2 - 3xy$

11. $x^3 - 2wy$

12. $w^3 - 2wx$

13. $\sqrt{2x}$

14. $\sqrt{z} - x^2$

15. $4wx^2 - 2wx$

16. $5wx^2 - 2w$

17. $4wxy - z^2$

18. $3xy^2 - wz$

Polynomials: Adding Polynomials

Name _____

Find each sum.

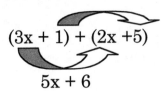

$(3x + 1) + (2x + 5)$

$5x + 6$

Add polynomials by adding like terms.

---

1. $(4n + 3) + (2n - 4)$

2. $(6x + 7) + (7x - 6)$

3. $\begin{array}{r} 3x^2 + 2x + 5 \\ + \underline{4x^2 + 5x + 6} \end{array}$

4. $\begin{array}{r} -y^2 + 3y - 5 \\ + \underline{y^2 + 6y + 9} \end{array}$

5. $(3m^2 + 2) + (m^2 - 1)$

6. $(x^2 - 5) + (3x^2 + 7)$

7. $(3 + 6m) + (4 - 2m)$

8. $(4x^2 - 4x) + (2x + 2)$

9. $\begin{array}{r} 6y^2 + y - 2 \\ + \underline{3y^2 - 2y + 4} \end{array}$

10. $\begin{array}{r} 6n^2 + 2n - 4 \\ + \underline{3n^2 - 2n + 4} \end{array}$

11. $(- x^2 + x) + (- x^2 - x - 2)$

12. $(3y^2 - 2y) + (-3y^2 + y)$

13. $(2y^2 + 5) + (3x - 4)$

14. $(6r + 4n) + (5r - 3n)$

15. $(-6m - 8n) + (5m + 8n)$

16. $(-21x + 6y) + (18x - 2y)$

17. $\begin{array}{r} -9x^2 + 3x - 20 \\ + \underline{6x^2 - 2x + 18} \end{array}$

18. $\begin{array}{r} 6ab^2 + 6ab + 6 \\ + \underline{3ab^2 - 4ab - 4} \end{array}$

Polynomials: Subtracting Polynomials

Name _____

Find each difference.

$(3x - 5) - (2x - 7)$ 1. Rewrite as an addition problem.
$3x - 5 + (^-2x + 7)$ 2. Add by combining like terms.
   $x + 2$

---

1. $(y + 5) - (3y - 6)$

2. $(4x + 2) - (2x + 3)$

3. $(9x + 5) - (x + 3)$

4. $(^-3x - 5) - (3x + 4)$

5. $(4n^2 + 3) - (2n^2 + 3)$

6. $(7x^2 + 3x + 1) - (^-x^2 - 4x)$

7. $(y^2 + 3y - 4) - (y^2 - 3y + 4)$

8. $(6x - 4) - (x + 6)$

9. $(^-6c - 5) - (^-c + 4)$

10. $(m^2 - 4m) - (2m^2 + 3m)$

11. $(^-y^2 + 3y - 4) - (2y^2 + 3)$

12. $(6 + 9y) - (4 + 8y)$

13. $(8x^2 + 3x - 5) - (6x^2 + 2x + 3)$

14. $(^-x - x^2) - (4 + 4x + 2x^2)$

15. $(9 + 3x^2 - 2x) - (10 - 4x^2 + 5x)$

16. $(6y^2 - 6y - 6) - (6y^2 + 6y + 6)$

THINK ABOUT IT!

17. Make up a polynomial subtraction problem with a difference of $^-2x^2 + 3x - 4$.

Polynomials: Multiplying Polynomials   Name _____

Part I: Multiplying monomials by polynomials. Find each product.

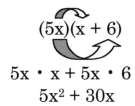

$(5x)(x + 6)$

$5x \cdot x + 5x \cdot 6$

$5x^2 + 30x$

Use the distributive property to multiply.

---

1. $3x(3x + 1)$

2. $y(4y + 3)$

3. $4x(x + 6)$

4. $(y + 5)(5y)$

5. $(x + 3)(x)$

6. $2p(p + 3)$

7. $(10x)(3x - 8)$

8. $6y(7y - 7)$

9. $6x(3 - 2x)$

10. $4x(6x - 3)$

11. $x(x^2 + 2x - 3)$

12. $3x(4x^2 + 2x + 6)$

Part II: Multiplying binomials by binomials. Find each product.

$(x - 3)(x + 4)$

$(x - 3)(x) + (x - 3)(4)$

$x^2 - 3x + 4x - 12$

$x^2 + x - 12$

Use the distributive twice to multiply.

13. $(x + 5)(x + 6)$

14. $(x + 3)(x - 5)$

15. $(2x - 2)(x + 3)$

16. $(x + 6)(2x - 7)$

17. $(3x + 1)(2x - 3)$

18. $(4x + 2)(5x - 1)$

19. $(3x + 5)(2x - 3)$

20. $(4x - 1)(4x - 1)$

21. $(2x + 2)(x - 8)$

Polynomials: Dividing Polynomials

Name _____

Part I: Dividing a monomial by a monomial. Find each quotient.

$$\frac{8x^3}{4x^2} = \frac{2 \cdot 2 \cdot 2 \cdot \cancel{x} \cdot \cancel{x} \cdot x}{\cancel{2} \cdot \cancel{2} \cdot \cancel{x} \cdot \cancel{x}}$$

$$\boxed{2x}$$

Divide by factoring the numerator and denominator and then canceling.

1. $\dfrac{42x^4}{14x^2}$

2. $\dfrac{9x^2}{3x}$

3. $\dfrac{3y^6}{15y^5}$

4. $\dfrac{-24x^3}{4x}$

5. $\dfrac{18x^5}{9x^3}$

6. $\dfrac{10y^4}{2y}$

7. $\dfrac{35m^6}{5m^2}$

8. $\dfrac{18c^5}{3c^2}$

9. $\dfrac{16y^3}{4y}$

Part II: Dividing a polynomial by a monomial. Find each quotient.

$$(9x^2 + 6x - 3) \div 3$$
$$\underline{9x^2 \div 3 + 6x \div 3 - 3 \div 3}$$
$$3x^2 + 2x - 1$$

Divide each term of the polynomial by the monomial.

10. $(2x^2 - 8x + 4) \div 2$

11. $(3x^2 + 21x + 6) \div 3$

12. $(5b^3 + 25b - 10) \div 5$

13. $(12x^3 - 4x + 8) \div 4$

14. $(2y^3 - 10y^2 + 6y) \div 2y$

15. $(9y^4 + 21y^3 - 6y) \div 3y$

16. $(14x^3 - 21x^2 + 7x) \div 7x$

17. $(16x^3 + 8x^2 + 12x) \div 2x$

ANSWER KEY

DIAGNOSTIC TEST

Page 5 1. D, 2. D, 3. A, 4. C, 5. A, 6. B, 7. D, 8. B 9. B, 10. A, 11. D, 12. B, 13. C, 14. A, 15. B, 16. D, 17. B, 18. C

Page 6 19. C, 20. D, 21. B, 22. C, 23. D, 24. B, 25. A, 26. D, 27. B, 28. B, 29. B, 30. B

Page 7 31. C, 32. D, 33. A, 34. C, 35. A, 36. A, 37. D, 38. B, 39. D, 40. C, 41. D, 42. B

Page 8 43. A, 44. D, 45. C, 46. A, 47. B, 48. B, 49. C, 50. B, 51. C, 52. C, 53. C

Page 9 1. 50, 2. 90, 3. 90, 4. 5,000, 5. 50, 6. 80, 7. 8, 8. 40, 9. 4,000, 10. 30, 11. 3,000, 12. 40, 13. 150, 14. 30, 15. 700, 16. 160, 17. 16,000, 18. 2,000, 19. 1,200, 20. 6,000, 21. 3,000, 22. 4,000, 23. 50,000, 24. 1,300

Page 10 1. $10 \cdot 10 \cdot 10 \cdot 10 \cdot 10$, 2. $12 \cdot 12 \cdot 12 \cdot 12$, 3. $2 \cdot 2 \cdot 2$, 4. $9 \cdot 9 \cdot 9 \cdot 9 \cdot 9$, 5. $3 \cdot 3 \cdot 3 \cdot 3 \cdot 3 \cdot 4 \cdot 4$, 6. $11 \cdot 11 \cdot 13 \cdot 13 \cdot 13$, 7. $n \cdot n \cdot n \cdot n$, 8. $x \cdot x \cdot x$, 9. $y \cdot y \cdot y \cdot y \cdot y \cdot z \cdot z$, 10. 3^3, 11. 12^2, 12. $6^2 \cdot 7^2$, 13. $2 \cdot 4^2$, 14. $8^3 \cdot 9^2$, 15. $5^2 \cdot 7^2 \cdot 8$, 16. $9ab^3$, 17. mn^2, 18. x^2y^2, 19. 100, 20. 2304, 21. 629, 22. 16,200, 23. 337, 24. 1,119,744

Page 11 1. 3^6, 2. 4^{10}, 3. 9^{10}, 4. 2^3, 5. 10^3, 6. 3^6, 7. 5^{13}, 8. 6^{11}, 9. 7^6, 10. 4, 11. 8^2, 12. 3, 13. a^6, 14. x^6, 15. y^8, 16. n^2, 17. b, 18. c, 19. a. 243, b. 150, c. 103 d. 64

Page 12 1. 7, 2. 47, 3. 77, 4. 9, 5. 1, 6. .5, 7. 14, 8. 9, 9. 145, 10. 35, 11. 29, 12. 16, 13. 27, 14. 2, 15. 5, 16. Sample Answer: $5 \cdot (4 + 3)$

Page 13 1. $6.00, 2. $40, 3. 2500, 4. 300, 5. a. $18 \div (3 + 6) = 2$, b. $15 \div (21 - 18) - 4 = 1$, c. $(54 + 24) \div 3 - 10 = 16$, d. $(24 \div 2) - (4 + 8) = 0$, 6. s^3

Page 14 1. 1000, 2. 8, 3. ⁻150, 4. 10, 5. 100, 6. ⁻20, 7. ⁻23, 8. 35, 9. 5, 10. 6, 11. 12, 12. 4, 13. 7, 14. 0, 15. 11, 16. 15, 17. 33, 18. 19, 19. 29, 20. 7

Page 15 1. ⁻7, 2. 36, 3. ⁻18, 4. ⁻33, 5. 16, 6. 18, 7. ⁻15, 8. ⁻1, 9. ⁻16, 10. ⁻41, 11. ⁻44, 12. 34, 13. 10, 14. 28, 15. ⁻31, 16. ⁻8, 17. ⁻9, 18. 0, 19. 0, 20. ⁻40,

Page 15 continued 21. 35, 22. a. 201, b. 0

Page 16 1. ⁻7, 2. ⁻15, 3. ⁻15, 4. ⁻7, 5. 40, 6. 9, 7. ⁻8, 8. ⁻28, 9. 62, 10. ⁻80, 11. 0, 12. 23, 13. 52, 14. 43, 15. ⁻77, 16. 19, 17. ⁻43, 18. ⁻53, 19. 0, 20. ⁻68, 21. 31, 22. Sample Answer: $x = 3, y = 2, z = 1$

Page 17 1. 15, 2. 16, 3. 24, 4. ⁻7, 5. ⁻81, 6. 54, 7. ⁻6, 8. ⁻56, 9. ⁻420, 10. ⁻4, 11. 7, 12. 0, 13. ⁻7, 14. ⁻1,242, 15. 1,248, 16. ⁻40, 17. 3,870, 18. 12, 19. 120, 20. ⁻39, 21. 168, 22. 1, 23. $x = ⁻2, y = ⁻3$ or $x = ⁻3, y = ⁻2$

Page 18 1. 85, 2. 13 yard gain, 3. 196, 4. 46°F, 5. 17th, 6. $130, 7. 23°F

Page 19 1. $8 + y$, 2. $m + 6$, 3. $r - 9$, 4. $6 - x$, 5. $n \div 9$, 6. $y - 5$, 7. $c \div 4$, 8. $12n$, 9. $t - 8$, 10. $11 + x$, 11. $n - 8 = 10$; 12. $9f + 6 = 24$, 13. $6x + 12 = 24$, 14. $7 + n \div 10 = 12$, 15. $h + $.07 = 2.09, 16. $4y - 8 = 32$

Page 20 1. 8, 2. 4, 3. 4, 4. 69, 5. 12, 6. 30, 7. 4, 8. 5, 9. 11, 10. 25, 11. 37, 12. 6, 13. 13, 14. 22, 15. 200, 16. 80, 17. 4, 18. 300, 19. $a = 9; b = 7$

Page 21 1. $c = 50$, 2. $y = 23$, 3. $x = 33$, 4. $r = 12$, 5. $x = 83$, 6. $n = 379$, 7. $c = 190$, 8. $n = 614$, 9. $n = 465$, 10. $n = ⁻91$, 11. $n = ⁻228$, 12. $s = 351$, 13. $x = ⁻73$, 14. $x = ⁻449$, 15. $x = ⁻35$, 16. $n = 131$, 17. $x = ⁻252$, 18. $m = ⁻15$, 19. $x = ⁻22$, 20. $x = 27$, 21. $c = ⁻64$, 22. $r = ⁻25$, 23. $X = 137$, 24. $R = ⁻19$

Page 22 1. $x = 18$, 2. $x = 23$, 3. $h = 26$, 4. $n = 630$, 5. $x = 8$, 6. $n = 414$, 7. $n = 247$, 8. $n = 704$, 9. $n = ⁻1,248$, 10. $y = 21$, 11. $x = ⁻62$, 12. $y = ⁻35$, 13. $n = ⁻434$, 14. $c = ⁻23$, 15. $y = 15$, 16. $n = 3,483$, 17. $h = ⁻10,625$, 18. $x = ⁻182$, 19. $n = 153$, 20. $x = ⁻6$, 21. $s = 8$, 22. $r = ⁻270$ 23. $c = ⁻12$, 24. $n = 399$

Page 23 1. $n = 18$, 2. $x = 40$, 3. $y = 3$, 4. $x = 21$, 5. $n = 9$, 6. $x = ⁻62$, 7. $x = 5$, 8. $x = 51$, 9. $r = 10$, 10. $h = 252$, 11. $n = ⁻54$, 12. $n = ⁻9$, 13. $n = ⁻5$, 14. $r = ⁻2$, 15. $e = 10$, 16. $x = 34$, 17. $x = ⁻185$, 18. $d = 103$, 19. a. $c = 7$, b. $y = ⁻5$, c. $x = 4$

Page 24 1. $x > 9$, 2. $c < 5$, 3. $x > 12$, 4. $m < 8$, 5. $c < 75$, 6. $n < 64$, 7. $x > ⁻5$, 8. $n > ⁻5$, 9. $x < ⁻5$, 10. $x > 8$, 11. $x < 19$, 12. $x \leq ⁻12$, 13. $c > ⁻6$, 14. $y \leq 3$

Page 24 continued 15. a > 2, 16. x < 9, 17. n ≥ 8, 18. a ≥ 4.5, 19. n < ⁻3.5, 20. x < 3, 21. y < 6, 22. 6

Page 25 1. $15, 2. $6.50, 3. Dr. Harrison is 3 years older than Dr. Schultz, 4. 3 x = 522; x = 174 5. y ÷ 8 = 5; y = 40, 6. It means y is between the values of 4 and 12. 7. true: a = 2 or 3; false: a = 0 or 1, 8. a. x can be any value, b. y = ⁻1, c. there is no solution.

Page 26 1. >, 2. >, 3. =, 4. <, 5. >, 6. <, 7. 3.069, 3.07, 3.7, 8. 4, 4.001, 4.01, 9. ⁻6.404, ⁻6.044, ⁻6.04, 10. ⁻0.101, ⁻0.011, ⁻0.001, 11. 20.003, 20.03, 20.303, 12. ⁻12.11, ⁻12.10, ⁻12.01

Page 27 1. 49, 2. 3, 3. 100, 4. 67, 5. 61, 6. 9, 7. 42, 8. 8, 9. 39, 10. 8.3, 11. 8.1, 12. 35.2, 13. 13.8, 14. 17.5, 15. 97.8, 16. 2.3, 17. 45.4, 18. 24.2, 19. 26.64, 20. 0.90, 21. 26.57, 22. 40.90, 23. 1.11, 24. 11.42

Page 28 1. 23.52, 2. 10.983, 3. 106.29, 4. 8.014, 5. 242.304, 6. 62.0966, 7. 0.378, 8. 21.22, 9. 3.88303, 10. 4.1, 11. 5.03, 12. ⁻37.12, 13. ⁻8.1, 14. 3.4, 15. 111.82, 16. 0, 17. ⁻4.48, 18. 0.91, 19. ⁻20.91, 20. ⁻5.7, 21. ⁻2.4

Page 29 1. 36.45, 2. 0.273, 3. 16.8096, 4. 19.722, 5. ⁻1.2832, 6. ⁻52.8, 7. 258.03, 8. ⁻7,520, 9. 3.4, 10. 3.6, 11. 11.3, 12. 16, 13. ⁻520, 14. 27, 15. ⁻1.3, 16. 15

Page 30 1. x = 43.44, 2. m = 501.5, 3. y = 18.3, 4. x = 20, 5. x = ⁻7, 6. a = .4, 7. a = 153.6, 8. x = ⁻19.09, 9. c = 46.8, 10. y = 17.28, 11. h = ⁻240, 12. t = ⁻40.82, 13. x = 113.1, 14. x = ⁻7.3, 15. m = ⁻15, 16. x = 3.53, 17. p = ⁻285, 18. r = 0, 19. x = ⁻15.6, 20. c = ⁻1.7, 21. t = ⁻90, 22. x = 6.4

Page 31 1. 420, 2. 4, 3. 6,700, 4. 914, 5. .08, 6. 3,000, 7. .42, 8. .0073, 9. 5, 10. 402, 11. 3.183, 12. 1,320, 13. 6,000, 14. 5,400, 15. 410,000, 16. 7.3, 17. .423, 18. .0032, 19. Sample Answer: When going from smaller units to larger units, divide by a power of ten. When going from larger to smaller units, multiply by a power of ten.

Page 32 1. 3.4×10^2, 2. 6.8×10^{-3}, 3. 5.4×10^3, 4. 7.18×10^6, 5. 2×10^{-5}, 6. 5.6×10^6, 7. 9.7×10^4, 8. 8.4×10^{-4}, 9. 2.3×10^4, 10. 320,000, 11. 5,203, 12. .00243, 13. 800000000, 14. .00000319,

Page 32 continued 15. .000283, 16. .000074, 17. 2,980, 18. 67, 19. 5.1×10^{-6}, 5.1×10^{-5}, 5.12×10^{-5}

Page 33 1. no; he will be overdrawn $110.43. 2. 3 for $1.10, 3. 25.3 miles, 4. 291 minutes, 5. six-pack, 6. 3.1536×10^7, 7. $96.40

Page 34 1. 9, 2. 64, 3. 225, 4. 169, 5. 81, 6. 49, 7. 196, 8. 400, 9. 100, 10. 256, 11. 121, 12. 289, 13. 625, 14. 324, 15. 144, 16. 5, 17. 8, 18. 9, 19. 11, 20. 3, 21. 10, 22. 20.518, 23. 9.592, 24. 29.12, 25. 12, 26. 20, 27. 25, 28. 9.327, 29. 7.55, 30. 11.18, 31. 2.8 cm

Page 35 1. 3, 6, 9, 12, 2. 7, 14, 21, 28, 3. 2, 4, 6, 8, 4. 8, 16, 24, 32, 5. 6, 12, 18, 24, 6. 9, 18, 27, 36, 7. 10, 20, 30, 40, 8. 20, 40, 60, 80, 9. 11, 22, 33, 44, 10. 25, 50, 75, 100, 11. 14, 28, 42, 56, 12. 15, 30, 45, 60, 13. 1, 2, 4, 8, 16, 14. 1, 2, 4, 7, 14, 28, 15. 1, 2, 3, 4, 6, 7, 12, 14, 21, 28, 42, 84, 16. 1, 2, 4, 5, 8, 10, 20, 40, 17. 1, 3, 5, 15, 25, 75, 18. 1, 2, 17, 34, 19. 1, 2, 3, 4, 5, 6, 10, 12, 15, 20, 30, 60, 20. 1, 3, 5, 7, 15, 21, 35, 105, 21. a. m = 24, b. y = 12, c. x = 8, d. c = 6; factors: 1, 2, 3, 4, 6, 8, 12, 24

Page 36 Note: Possible answers are given for factors of composite numbers. 1. C; 2, 2. C; 7, 3. C; 6, 4. P, 5. P, 6. C; 9, 7. P, 8. C; 5, 9. C; 6, 10. C; 8, 11. C; 5, 12. P, 13. C; 4, 15. C; 3, 16. P, 17. C; 9, 18. P, 19. C; 10, 20. C; 11, 21. C; 9, 22. C; 10, 23. C; 2, 24. C; 2, 25. P, 26. P, 27. C; 6, 28. C; 13, 29. x = 6

Page 37

1. 36 $2^2 * 3^2$
   6 6
   2 3 2 3

2. 90 $2 * 3^2 * 5$
   9 10
   3 3 2 5

3. 40 $2^3 * 5$
   5 8
   2 4
   2 2

4. 85 $5 * 17$
   5 17

5. 51 $3 * 17$
   3 17

6. 225 $3^2 * 5^2$
   25 9
   5 5 3 3

7. a. n = 2
   b. x = 2
   c. n = 3

92

Page 38 1. 8, 2. 1, 3. 8, 4. 12, 5. 1, 6. 18, 7. 15, 8. 2, 9. 15, 10. 6, 11. 4, 12. 6, 13. a. $3xy^2$, b. $2z^2$, c. xyz

Page 39 1. 72, 2. 35, 3. 210, 4. 120, 5. 240, 6. 900, 7. 30, 8. 3750, 9. 40, 10. 66, 11. 84, 12. 180, 13. 30y, 14. 24xy, 15. 20xy

Page 40 1. 3 rows of 6, 2 rows of 9, 6 rows of 3, 2. 3 and 5, 11 and 13, 17 and 19, 29 and 31, 41 and 43, 3. an error message; you cannot take the square root of a negative number because the roots have to be exactly the same; 4. 3.5 cm, 5. 20: 1, 2, 4, 5, 10, 20; 30: 1, 2, 3, 5, 6, 10, 15, 30; LCM of 20 and 30 = 60: 1, 2, 3, 4, 5, 6, 10, 12, 15, 20, 30, 60. All the factors of 20 and 30 are also factors of 60. 6. a. yes, b. no; GCF = 5; c. no; GCF = 7, 7. a. x = 3; b. x= 2

Page 41 1. $\frac{10}{13}$, 2. $\frac{11}{17}$, 3. $\frac{1}{9}$, 4. $\frac{5}{8}$, 5. $\frac{8}{9}$, 6. $\frac{3}{4}$, 7. $\frac{1}{2}$, 8. $\frac{7}{10}$, 9. $\frac{9}{11}$, 10. $\frac{7}{15}$, 11. $\frac{5}{6}$, 12. $\frac{7}{9}$, 13. $\frac{3}{5}$, 14. $\frac{3}{17}$, 15. $\frac{6}{7}$, 16. $\frac{3}{10}$, 17. $\frac{7}{15}$, 18. $\frac{55}{78}$, 19. $\frac{21}{32}$, 20. $\frac{7}{20}$, 21. $\frac{17}{19}$, 22. $\frac{2}{3}$, 23. $\frac{71}{82}$, 24. $\frac{7}{11}$, 25. a. $\frac{3}{4}$, b. $\frac{y}{2}$, c. $\frac{y}{2}$

Page 42 1. $6\frac{3}{4}$, 2. $2\frac{5}{11}$, 3. $4\frac{1}{2}$, 4. $9\frac{1}{4}$, 5. 17, 6. 6, 7. $5\frac{1}{7}$, 8. $1\frac{3}{4}$, 9. 1, 10. $4\frac{3}{8}$, 11. $15\frac{3}{5}$, 12. 3, 13. $4\frac{2}{5}$, 14. $3\frac{2}{7}$, 15. $14\frac{1}{3}$, 16. $\frac{47}{8}$, 17. $\frac{26}{3}$, 18. $\frac{95}{11}$, 19. $\frac{51}{15}$, 20. $\frac{31}{8}$, 21. $\frac{191}{40}$, 22. $\frac{19}{5}$, 23. $\frac{19}{8}$, 24. $\frac{33}{5}$, 25. $\frac{207}{51}$, 26. $\frac{123}{11}$, 27. $\frac{51}{8}$, 28. $\frac{44}{7}$, 29. $\frac{60}{7}$, 30. $\frac{89}{10}$

Page 43 1. <, 2. <, 3. <, 4. <, 5. <, 6. <, 7. >, 8. <, 9. >, 10. >, 11. <, 12. >, 13. <, 14. <, 15. <, 16. $\frac{-7}{8}$, $\frac{-3}{4}$, $\frac{-5}{7}$, 17. $\frac{2}{9}$, $\frac{2}{7}$, $\frac{2}{5}$, $\frac{2}{3}$, 18. $\frac{3}{8}$, $\frac{3}{7}$, $\frac{5}{6}$, $\frac{6}{7}$, 19. $\frac{-3}{4}$, $\frac{-2}{3}$, $\frac{-5}{12}$, $\frac{-6}{15}$, 20. $\frac{4}{5}$, $\frac{5}{6}$, $\frac{8}{9}$, $\frac{9}{10}$, 21. $\frac{1}{5}$, $\frac{1}{4}$, $\frac{1}{3}$

Page 44 1. $\frac{11}{12}$, 2. $\frac{1}{4}$, 3. $\frac{7}{12}$, 4. $1\frac{7}{12}$, 5. $-1\frac{3}{20}$, 6. $\frac{5}{28}$, 7. $1\frac{1}{24}$, 8. $-1\frac{1}{5}$, 9. $\frac{3}{4}$, 10. $1\frac{5}{21}$, 11. $\frac{7}{24}$, 12. $\frac{1}{9}$

Page 44 13. $1\frac{8}{15}$, 14. $\frac{17}{40}$, 15. $1\frac{17}{24}$, 16. $-1\frac{1}{3}$

Page 45 1. $2\frac{1}{4}$, 2. $\frac{1}{2}$, 3. $-1\frac{3}{4}$, 4. $1\frac{3}{4}$, 5. $3\frac{7}{12}$, 6. $\frac{4}{5}$, 7. $3\frac{7}{30}$, 8. $-10\frac{3}{4}$, 9. $24\frac{3}{40}$, 10. $2\frac{11}{20}$, 11. $11\frac{1}{36}$, 12. $3\frac{1}{18}$, 13. $6\frac{7}{45}$, 14. 0, 15. $16\frac{1}{40}$, 16. $\frac{-11}{14}$, 17. $6\frac{7}{45}$, 18. $1\frac{1}{15}$

Page 46 1. x = 4, 2. x = $\frac{-19}{20}$, 3. m = $1\frac{1}{60}$, 4. n = $4\frac{1}{8}$, 5. x = 0, 6. c = $\frac{-9}{50}$, 7. y = $\frac{1}{2}$, 8. x = $\frac{7}{12}$, 9. y = $-1\frac{19}{60}$, 10. x = $-1\frac{19}{30}$, 11. n = $\frac{1}{4}$, 12. x = $\frac{1}{35}$, 13. y = $\frac{9}{20}$, 14. w = $-1\frac{5}{72}$, 15. x = $\frac{3}{14}$

Page 47 1. $\frac{9}{20}$, 2. $\frac{5}{32}$, 3. $14\frac{5}{8}$, 4. 4, 5. $23\frac{4}{7}$, 6. -18, 7. $2\frac{1}{2}$, 8. $-18\frac{3}{8}$, 9. 1, 10. $5\frac{5}{14}$, 11. 14, 12. $-12\frac{8}{9}$, 13. $-1\frac{2}{5}$, 14. $-10\frac{1}{3}$, 15. $-12\frac{3}{5}$, 16. 30, 17. $\frac{5}{9}$, 18. $2\frac{1}{2}$, 19. $\frac{-3}{35}$, 20. $144\frac{5}{6}$, 21. a. $2\frac{1}{4}$, b. $-3\frac{3}{4}$, c. $-11\frac{23}{32}$

Page 48 1. 3, 2. $\frac{-3}{16}$, 3. $\frac{-1}{12}$, 4. $3\frac{1}{5}$, 5. 8, 6. $-1\frac{5}{16}$, 7. $\frac{-16}{25}$, 8. $-3\frac{1}{3}$, 9. $\frac{-8}{11}$, 10. $-3\frac{3}{4}$, 11. $-2\frac{5}{8}$, 12. $-6\frac{1}{4}$, 13. -6, 14. -10, 15. $2\frac{1}{2}$, 16. $9\frac{3}{5}$, 17. -14, 18. $-1\frac{7}{16}$

Page 49 1. $-3\frac{1}{3}$, 2. y = $\frac{1}{10}$, 3. m = $4\frac{4}{5}$, 4. y = $7\frac{1}{5}$, 5. y = 4, 6. x = $1\frac{1}{9}$, 7. x = -8, 8. b = $\frac{-3}{35}$, 9. x = $\frac{-18}{25}$, 10. m = $1\frac{1}{2}$, 11. x = $27\frac{3}{5}$, 12. m = $-1\frac{29}{85}$, 13. x = $\frac{33}{46}$, 14. x = $\frac{-2}{3}$, 15. n = $\frac{-3}{25}$

Page 50 1. $\frac{1}{5}$, 2. $\frac{17}{6}$, 3. $\frac{4}{7}$, 4. $\frac{3}{5}$, 5. $\frac{5}{13}$, 6. $\frac{1}{5}$,

Page 50 continued 7. $\frac{2}{5}$, 8. $\frac{9}{16}$, 9. $\frac{1}{9}$, 10. no, 11. no, 12. no, 13. yes, 14. yes, 15. no, 16. no, 17. no, 18. no, 19. yes, 20. no, 21. yes

Page 51 1. n = 12, 2. n = 5.6, 3. n = 17, 4. x = 18, 5. n = 14, 6. x = 16, 7. x = 14, 8. n = 2, 9. x = 4, 10. y = 6, 11. n = 49, 12. h = 40; 13. n = 3, 14. x = 81, 15. c = 3.5, 16. x = 1.5, 17. n = 4.5, 18. p = 16; 19. w = 4, 20. c = 30, 21. n = 56, 22. n = 63, 23. n = 18, 24. n = 3, 25. a. 29, b. 17, c. 3

Page 52 1. 16 people per car, 2. 140 tickets per hour, 3. 25 miles per gallon, 4. $.07 per egg, 5. 135 miles per day, 6. 4.5 people per car, 7. 3 pounds in 1 week, 8. $1.99 per pound, 9. $214 per day, 10. 5 people per row, 11. $7.25 per tape, 12. 63 miles per hour, 13. $48 per outfit, 14. 25 students per teacher, 15. 3600 words per hour, 16. $6.00 per tape, 17. 48 minutes

Page 53 1. 88%, 2. 53%, 3. 25%, 4. 36%, 5. 1%, 6. 99%, 7. 27%, 8. 63%, 9. 15%, 10. 43%, 11. 21%, 12. 71%, 13. $\frac{3}{5}$, 14. $\frac{17}{100}$, 15. $\frac{1}{20}$, 16. $\frac{1}{5}$, 17. 1, 18. $\frac{7}{20}$, 19. $\frac{2}{25}$, 20. $\frac{23}{400}$, 21. $\frac{19}{200}$, 22. $\frac{63}{100}$, 23. $\frac{92}{100}$, 24. $\frac{44}{100}$, 25. a. ◄———► b. ▦ c. ■□□□□ d. ◉

Page 54 1. 50.8%, 2. 1%, 3. 40%, 4. 55%, 5. 240%, 6. 7%, 7. 2.3%, 8. 75%, 9. 30%, 10. 428%, 11. 24.5%, 12. .6%, 13. .72, 14. .04, 15. .17, 16. .9, 17. .121, 18. 2, 19. .25, 20. .0025, 21. .0075, 22. .105, 23. 4.5, 24. .8

Page 55 1. 684, 2. 14,000, 3. $47.50, 4. 20, 5. 4½ cups, 6. 86%, 7. 4

Page 56 1. 20%, 2. 2.2, 3. .75, 4. 2.4, 5. 60, 6. 54.3, 7. 50%, 8. 45, 9. 50, 10. 1, 11. 130 %, 12. 2.7, 13. 123.6, 14. 100%

Page 57 1. $1134, 2. $61.20, 3. $24, 4. $4.92, 5. $60.13, 6. $171, 7. $472.50, 8. $26.25, 9. $7.19, 10. $205

Page 58 **I** = **i**ncrease; **D** = **d**ecrease 1. I, 20%, 2. D, 50%, 3. D, 15%, 4. I, 30%, 5. I, 56%, 6. I, 18%, 7. D, 11%, 8. D, 44%, 9. I, 16%, 10. I, 35%, 11. I, 12%, 12. I, 9%, 13. D, 36%, 14. D, 63%, 15. I, 33% 16. $14.65

Page 59 1. 195, 2. 71%, 3. 73%, 4. yes, 5. 67%, 6. $900, 7. 32.5%, 8. 15%

Page 60 1. 21 m, 2. 10.5 m, 3. 69 cm, 4. 36 in, 5. 10 m, 6. 9.5 in, 7. 8.3 in, 8. 7.5 in

Page 61 1. right, 2. straight, 3. acute, 4. obtuse, 5. acute, 6. right, 7. x + 123 = 180; x = 57°, 8. x + 57 = 90; x = 33°, 9. x + 43 = 90; x = 47°, 10. x + 45 = 180; x = 135°, 11. x + 19 = 90; x = 71°, 12. x + 93 = 180; x = 87°, 13. x + 110 = 180; x = 70°, 14. x + 38 = 90; x = 52°

Page 62 1. equilateral acute, 2. scalene obtuse, 3. scalene right, 4. isosceles obtuse, 5. x = 70°, 6. x = 32°, 7. x = 50°, 8. x = 60°

Page 63 1. 15 m, 2. 5 cm, 3. 18.9 in, 4. 17 ft, 5. 25.1 m, 6. 13 in, 7. 23.3 ft, 8. 11.4 cm, 9. 25 ft, 10. 16.2 mm, 11. 41 in, 12. 14.2 ft, 13. yes, 14. no, 15. yes, 16. no, 17. yes, 18. yes, 19. yes, 20. no, 21. no

Page 64 1. C = 157 m, A = 1962.5 m², 2. C = 15.07 ft, A = 18.09 ft², 3. C = 62.8 m, A = 314 m², 4. C = 24.49 in, A = 47.76 in², 5. C = 10.99 yd, A = 9.62 yd², 6. C = 188.4 m, A = 2826 m², 7. a. d = 13 cm, r = 6.5 cm, b. d = 4.4 in, r = 2.2 in, c. d = 24 cm, r = 12 cm

Page 65 1. 84 m², 2. 87.84 m², 3. 141.12 m², 4. 684 m², 5. 11.34 m², 6. 22.04 m², 7. 159.6 m², 8. 115 m², 9. 442.32 m², 10. 69.87 m², 11. 105 cm²

Page 66 1. 67.7 m², 2. 6 ft², 3. 2.0 cm², 4. 7.5 cm², 5. 65 in², 6. 75.7 m², 7. 41.4 m², 8. 1.3 ft²

Page 67 1. 508.68 m³, 2. 231 in³, 3. 540 m³, 4. 9847.04 cm³, 5. 18.75 cm³, 6. 5595.48 in³, 7. 706.5 m³, 8. 19.32 cm³, 9. 11,232 in³

Page 68 1. 234.7 cm³, 2. 1657.9 m³, 3. 165 m³, 4. 1356.5 m³, 5. 506.7 m³, 6. 251.2 m³, 7. 158.3 cm³, 8. 261.7 m³, 9. a. 4500 π, b. 2304 π, c. 972 π, d. 288 π

Page 69 1. 326.56 m^2, 2. 234 cm^2, 3. 255 in^2, 4. 452.16 m^2, 5. 382 in^2, 6. 3,165.12 cm^2, 7. 602.88 in^2, 8. 150 m^2, 9. Possible Answer: Find the area of one of the circular bases then multiply by two. Add that to π d • h which is the area of the middle part.

Page 70 1. $2,146.67, 2. 6 m, 3. 16 cm, 4. 20 ft, 5. 6 ft, 6. $9,660.00, 7. 18 m, 8. x + 68 + 34 = 180; x = 78°; acute, 9. it doubles

Page 71 1. 54, 2. 28, 3. 90, 4. 12, 5. 16, 6. 60, 7. 36

Page 72 1. 6, 2. 5,040, 3. 24, 4. 120, 5. 24, 6. 6, 7. 3,628,800, 8. 39,916,800, 9. 6, 10. 120, 11. 720, 12. 6, 13. 151,200

Page 73 1. 6, 2. 5, 3. 28, 4. 35, 5. 1140, 6. 120, 7. 3, 8. 33649, 9. 45, 10. 10, 11. 66, 12. 5005, 13. 126, 14. 15, 15. 66

Page 74 1. $\frac{1}{16}$, 2. $\frac{1}{169}$, 3. $\frac{1}{169}$, 4. $\frac{1}{4}$, 5. $\frac{1}{4}$, 6. $\frac{1}{2704}$, 7. $\frac{1}{16}$, 8. $\frac{1}{676}$, 9. $\frac{4}{91}$, 10. $\frac{4}{91}$, 11. $\frac{5}{91}$, 12. $\frac{5}{182}$, 13. $\frac{1}{91}$, 14. $\frac{3}{182}$

Page 75 1. 36, 2. 1,024, 3. 40,320, 4. a. $\frac{5}{18}$, b. $\frac{5}{9}$, c. $\frac{17}{18}$, d. $\frac{1}{9}$, 5. $\frac{5}{39}$, 6. 42

Page 76

1.		2.	
1	12446	5	18
2	2	6	235
3	25	7	1299
4	56	8	39

3.		4.	
	089		73
1	0789	8	5677
2	02267	9	01255
3	1		

5. 63, 100, 6. 20, 7. in the 80's, 8. 87.5, 9. 86.65, 10. Most of the test scores were above average.

Page 77 1. 20, 2. 80-89, 3. 16 out of 20 scored 70 or above so most of the students studied. 4. indiv. test scores, 5. 80-89, 6. no, 7. a tropical climate

Page 78 1.

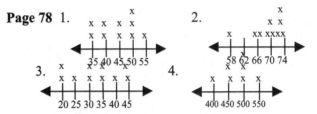

5. 6 ft 7 in, 5 ft 11 in, 6. 72 in & 74 in, 7. Shooter's, 8. They both use columns to display data; you can see individual data in a line plot.

Page 79
1. A (-7, -8), 2. B (2, -2), 3. C (0, 6), 4. D (–8, 8)
5. E (6, 0), 6. F (4, 7)

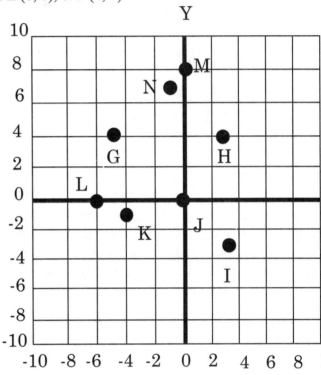

15. See student graphs; the graph is symmetrical; they are reflections.

Page 80
1. x ≥ -2;   2. x < -1;
3. x > 4;   4. x ≤ -2;
5. x ≤ -5;   6. x > 10;
7. x ≤ -4;   8. x > 1;
9. a.   b.

Page 81

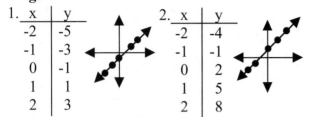

x	y
-2	-5
-1	-3
0	-1
1	1
2	3

2.

x	y
-2	-4
-1	-1
0	2
1	5
2	8

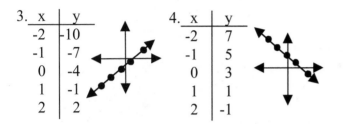

3.

x	y
-2	-10
-1	-7
0	-4
1	-1
2	2

4.

x	y
-2	7
-1	5
0	3
1	1
2	-1

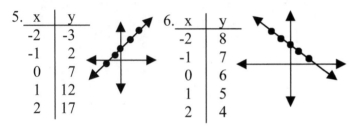

5.

x	y
-2	-3
-1	2
0	7
1	12
2	17

6.

x	y
-2	8
-1	7
0	6
1	5
2	4

7. x = 6, y = 7

Page 82 1. y = 319, 2. x = ⁻23, 3. x = 86, 4. y = 12, 5. n = 5, 6. y = 25, 7. x = 2, 8. y = 20, 9. b = ⁻3, 10. b = ⁻3, 11. x = ⁻9, 12. a = ⁻5, 13. x = ⁻12, 14. y = 23, 15. x = ⁻13, 16. y = ⁻21, 17. x = ⁻6, 18. x = ⁻12, 19. y = 16, 20. x = 9, 21. n = 6.5

Page 83 1. x = 2, 2. n = ⁻4, 3. y = 16, 4. x = ⁻11, 5. n = 12, 6. x = 1.6, 7. y = 20, 8. n = 4, 9. y = 15, 10. x = 2, 11. n = 7, 12. b = 5

Page 84 1. x = 9, 2. n = ⁻4, 3. x = ⁻12, 4. y = ⁻7, 5. b = 2, 6. n = ⁻3, 7. n = 2, 8. x = 1, 9. n = 8, 10. x = 12, 11. x = ⁻12, 12. y = 10, 13. x = 6, 14. y = ⁻5, 15. x = ⁻3, 16. x = 4, 17. n = 11, 18. a = 5, 19. x = 2, 20. y = 2, 21. a = ⁻2

Page 85 1. a < ⁻6, 2. x ≤ 8, 3. y < 6, 4. x < 2, 5. x ≥ ⁻3, 6. c ≤ 3, 7. h ≥ 75, 8. x < ⁻93, 9. y ≥ ⁻46, 10. n ≤ 6, 11. m > ⁻3, 12. k ≤ ⁻12, 13. x > ⁻4, 14. y < ⁻3, 15. y > 2, 16. a ≤ 4, 17. y ≥ 3, 18. x > 1, 19. n < .25, 20. x > ⁻5, 21. y > ⁻5

Page 86 1. trinomial, 2. monomial, 3. binomial, 4. trinomial, 5. monomial, 6. trinomial, 7. binomial, 8. trinomial, 9. monomial, 10. 22, 11. 2, 12. 3, 13. 2, 14. ⁻2, 15. ⁻12, 16. ⁻18, 17. 8, 18. 58

Page 87 1. 6n − 1, 2. 13x + 1, 3. 7x² + 7x + 11, 4. 9y + 4, 5. 4m² + 1, 6. 4x² + 2, 7. 7 + 4m, 8. 4x² − 2x + 2, 9. 9y² − y + 2, 10. 9n², 11. ⁻2x² − 2, 12. ⁻y, 13. 2y² + 3x + 1, 14. 11r + n, 15. ⁻m, 16. ⁻3x + 4y, 17. ⁻3x² + x − 2, 18. 9ab² + 2ab + 2

Page 88 1. ⁻2y + 11, 2. 2x − 1, 3. 8x + 2, 4. ⁻6x − 9, 5. 2n², 6. 8x² + 7x + 1, 7. 6y − 8, 8. 5x − 10, 9. ⁻5c − 9, 10. ⁻m² − 7m, 11. ⁻3y² + 3y − 7, 12. 2 + y, 13. 2x² + x − 8, 14. ⁻3x² − 5x − 4, 15. ⁻1 + 7x² − 7x, 16. ⁻12y − 12, 17. Answers will vary for this problem: (6x² + 2x − 7) − (8x² − x − 3)

Page 89 1. 9x² + 3x, 2. 4y² + 3y, 3. 4x² + 24x, 4. 5y² + 25y, 5. x² + 3x, 6. 2p² + 6p, 7. 30x² − 80x, 8. 42y² − 42y, 9. 18x − 12x², 10. 24x² ⁻12x, 11. x³ + 2x² − 3x, 12. 12x³ + 6x² + 18x, 13. x² + 11x + 30, 14. x² − 2x − 15, 15. 2x² + 4x − 6, 16. 2x² + 5x − 42, 17. 6x² − 7x − 3, 18. 20x² + 6x − 2, 19. 6x² + x − 15, 20. 16x² − 8x + 1, 21. 2x² − 14x − 16

Page 90 1. 3x², 2. 3x, 3. $\frac{1}{5}$y, 4. ⁻6x², 5. 2x², 6. 5y³, 7. 7m⁴, 8. 6c³, 9. 4y², 10. x² − 4x + 2, 11. x² + 7x + 2, 12. b³ + 5b − 2, 13. 3x³ − x + 2, 14. y² − 5y + 3 15. 3y³ + 7y² − 2, 16. 2x² − 3x + 1, 17. 8x² + 4x + 6